The Post-Critical Kant

"In this book Bryan Hall has followed up his articles on Kant's post-Critical writings to focus on the familiar puzzle of a supposed 'gap' in the Critical work, which the *Opus Postumum* was designed to bridge. His book promises to throw real light on Kant's later work and its controversial relation to the earlier *Critiques.*"

—Graham Bird, *Emeritus Professor of Philosophy, Manchester University, UK*

"After the publication of the B edition of the *Critique of Pure Reason* in 1787, Kant gradually came to the realization that there is a fundamental "gap" in his explanation of transcendental idealism; and in his final unfinished "Transition" project of the late 1790s and early 1800s, now known as the *Opus postumum,* he tried desperately to close this gap. Bryan Hall's *The Post-Critical Kant* explores this profound problem in Kant's metaphysics and epistemology in a highly thoroughgoing, incisive, and compelling way that will at the very least significantly inflect, and perhaps even drastically change, our conception of the Critical philosophy itself."

—Robert A. Hanna, *Professor of Philosophy, University of Colorado, USA*

In this book, Bryan Wesley Hall breaks new ground in Kant scholarship, exploring the gap in Kant's Critical philosophy in relation to his post-Critical work by turning to Kant's final, unpublished work, the so-called *Opus Postumum.* Although Kant considered this project to be the "keystone" of his philosophical efforts, it has been largely neglected by scholars. Hall argues that only by understanding the *Opus Postumum* can we fully comprehend both Kant's mature view as well as his Critical project.

In letters from 1798, Kant claims to have discovered a "gap" in the Critical philosophy that requires effecting a "transition from the metaphysical foundations of natural science to physics"; unfortunately, Kant does not make clear exactly what this gap is or how the transition is supposed to fill the gap. To resolve these issues, Hall draws on the *Opus Postumum,* arguing that Kant's transition project can solve certain perennial problems with the Critical philosophy. This volume provides a powerful alternative to all current interpretations of the *Opus Postumum,* arguing that Kant's transition project is best seen as the post-Critical culmination of his Critical philosophy. Hall carefully examines the deep connections between the *Opus Postumum* and the view Kant develops in the *Critique of Pure Reason,* to suggest that properly understanding the post-Critical Kant will significantly revise our view of Kant's Critical period.

Bryan Wesley Hall is Associate Professor of Philosophy at Indiana University Southeast, USA.

Routledge Studies in Eighteenth-Century Philosophy

1 **Naturalization of the Soul**
Self and Personal Identity in the Eighteenth Century
Raymond Martin and John Barresi

2 **Hume's Aesthetic Theory**
Taste and Sentiment
Dabney Townsend

3 **Thomas Reid and Scepticism**
His Reliabilist Response
Philip de Bary

4 **Hume's Philosophy of the Self**
A E Pitson

5 **Hume, Reason and Morality**
A Legacy of Contradiction
Sophie Botros

6 **Kant's Theory of the Self**
Arthur Melnick

7 **Aesthetics and Morals in the Philosophy of David Hume**
Timothy M. Costelloe

8 **Hume's Difficulty**
Time and Identity in the Treatise
Donald L.M. Baxter

9 **Kant and the Cultivation of Virtue**
Chris W. Surprenant

10 **The Post-Critical Kant**
Understanding the Critical Philosophy through the *Opus postumum*
Bryan Wesley Hall

The Post-Critical Kant

Understanding the Critical Philosophy through the *Opus postumum*

Bryan Wesley Hall

Routledge
Taylor & Francis Group
NEW YORK AND LONDON

First published 2015
by Routledge
711 Third Avenue, New York, NY 10017

and by Routledge
2 Park Square, Milton Park, Abingdon, Oxon OX14 4RN

First issued in paperback 2017

Routledge is an imprint of the Taylor & Francis Group, an informa business

Library of Congress Cataloging-in-Publication Data
Hall, Bryan, 1977–
Understanding the critical philosophy through the *Opus postumum* / Bryan Wesley Hall. — 1 [edition].
pages cm. — (Routledge studies in eighteenth-century philosophy ; 10)
Includes bibliographical references and index.
1. Kant, Immanuel, 1724–1804. *Opus postumum*. 2. Transcendentalism. 3. Physics—Philosophy. 4. Kant, Immanuel, 1724–1804. I. Title.
B2794.O63H35 2014
193—dc23
2014017007

ISBN 13: 978-1-138-09864-0 (pbk)
ISBN 13: 978-1-138-80214-8 (hbk)

Typeset in Sabon
by Apex CoVantage, LLC

Contents

Acknowledgements

The seeds for this project were first sown over ten years ago when I was working on my dissertation, first in Marburg, Germany, and later in Boulder, Colorado. Consequently, my first and most heartfelt thanks must go to my two *Doktorväter*: Burkhard Tuschling and Robert Hanna. Whereas Burkhard fostered the first sprouts of this project with a careful hand and a generous heart, Bob saw it through to fruition with constant encouragement and unyielding support over the course of many years. Since Burkhard is no longer with us, I would like to dedicate this book to him. He is (and will be) dearly missed.

Although Burkhard and Bob provided much of the philosophical support I needed to complete this project, it would not have been possible were it not for significant financial support as well. I would like to thank the German-American Fulbright Commission for the 2003–2004 junior award that allowed me to work on my dissertation in Marburg. I would also like to thank Indiana University Southeast for giving me a summer research fellowship in 2006 to start work on the manuscript itself. A senior award from the Romanian-American Fulbright Commission permitted me to complete a draft of the manuscript in 2010–2011. Thanks also to the Philosophy faculty at the University of Bucharest for providing such a supportive environment. Finally, I was able to complete the book thanks to a sabbatical from Indiana University Southeast in spring 2014.

Many people have contributed to making this a much better book than it would have otherwise been. In particular, I would like to thank Jeffrey Edwards, Rae Langton, and Walter Ott, who all offered insightful comments on this book either in part or whole and whether they knew it or not. Thanks also to the two anonymous referees at Routledge whose comments proved tremendously helpful as I revised the manuscript for publication. Over the years, I have presented parts of this book to many audiences both in the U.S. and abroad. Although I cannot thank them individually here, their comments were most welcome and I appreciated their willingness to wade into the dark waters of the *Opus postumum* with me.

Of the many secondary sources I read during my research, I was most fortunate to read Giovanni Pietro Basile's new book on the *Opus postumum* before my own book went to press. It is by far the most complete account of the secondary literature on the *Opus postumum* in print. For someone whose only languages are English and German, his book introduced me to all the fine work being done in other European languages. I have cited some of this literature in my manuscript with a reference to Basile's discussion for those who face similar linguistic challenges.

Some chapters of this book are based on previously published material. Chapter One is based on my article, "A Dilemma for Kant's Theory of Substance," *British Journal for the History of Philosophy* 19, no. 1 (2011): 79–109. Chapter Three expands on my article, "A Reconstruction of Kant's Ether Deduction in *Übergang* 11," *British Journal for the History of Philosophy* 14, no. 4 (2006): 719–746. I would like to thank the *British Journal for the History of Philosophy* as well as its publisher, Taylor and Francis, for permission to reuse material from both articles. Chapter Four develops my article, "*Effecting a Transition*: How to Fill the Gap in Kant's System of Critical Philosophy," *Kant-Studien* 100, no. 2 (2009): 187–211. I would like to thank *Kant-Studien* and its publisher, De Gruyter, for permission to reuse material from this article. Finally, Chapter Five incorporates parts of my article, "Appearances and the Problem of Affection in Kant," *Kantian Review* 14, no. 2 (2010): 38–66. Thanks to *Kantian Review* and its publisher, Cambridge University Press, for permission to reuse material from this article.

Unless otherwise noted, translated quotes from Kant's works are taken from the Cambridge University Press series of translations. These quotes were reprinted with the permission of Cambridge University Press.

Translations and Abbreviations

With the exception of the *Critique of Pure Reason*, references to Kant's works cite the *Akademie* edition of *Kants gesammelte Schriften*. I first cite volume and then page number. References to the *Critique of Pure Reason* use the standard A-edition (1781) and B-edition (1787) citation format. For convenience, when I quote Kant's works, I use parenthetical in-text citations. Unless otherwise noted, all translations come from the Cambridge University Press series of Kant's works. When a translation is my own, I say so in an endnote and, within the parenthetical citation, offer line numbers in addition to volume and page to aid readers in locating the passage. Below I provide bibliographical information for the Cambridge translations I used. When I have cited a work more than once, I also provide an abbreviation for the work.

Abbreviation	*Translation*
AK	*Kants gesammelte Schriften* (29 vols.). Berlin: De Gruyter, 1902–.
	Concerning the Ultimate Ground of the Differentiation of Directions in Space. In *Theoretical Philosophy 1755–1770*, translated by David Walford and Ralf Meerbote. Cambridge: Cambridge University Press, 1992.
CJ	*Critique of the Power of Judgment,* translated by Paul Guyer and Eric Matthews. Cambridge: Cambridge University Press, 2000.
Correspondence	*Correspondence,* translated by Arnulf Zweig. Cambridge: Cambridge University Press, 1999.
	Critique of Practical Reason, translated by Mary Gregor. In *Practical Philosophy,* edited by Mary Gregor. Cambridge: Cambridge University Press, 1996.
CPR	*Critique of Pure Reason,* translated by Paul Guyer and Allen Wood. Cambridge: Cambridge University Press, 1997.

MFNS	*Metaphysical Foundations of Natural Science,* translated by Michael Friedman. In *Theoretical Philosophy after 1781,* edited by Henry Allison and Peter Heath. Cambridge: Cambridge University Press, 2002.
MM	*Metaphysics of Morals,* translated by Mary Gregor. In *Practical Philosophy,* edited by Mary Gregor. Cambridge: Cambridge University Press, 1996.
OP	*Opus postumum,* translated by Eckart Förster and Michael Rosen. Cambridge: Cambridge University Press, 1993.
Prolegomena	*Prolegomena to any Future Metaphysics Metaphysical Foundations of Natural Science,* translated by Gary Hatfield. In *Theoretical Philosophy after 1781,* edited by Henry Allison and Peter Heath. Cambridge: Cambridge University Press, 2002.

Introduction

> Thus with this I bring my entire critical enterprise to an end. I shall proceed without hindrance to the doctrinal part, in order, if possible, to win yet from my increasing age some time still favorable to that. It is self-evident that there will be no special part for the power of judgment in that, since in regard to that critique serves instead of theory; rather, in accordance with the division of philosophy into theoretical and practical parts, and the division of pure philosophy into the very same parts, the metaphysics of nature and of morals will constitute that enterprise.
>
> —Kant, Preface to the *Critique of the Power of Judgment* from 1790 (*CJ* 5:170)

> I see before me the unpaid bill of my uncompleted philosophy, even while I am aware that philosophy, both as regards its means and its ends, is capable of completion. It is a pain like that of Tantalus though not a hopeless pain. The project on which I am now working concerns the 'Transition from the metaphysical foundations of natural science to physics.' It must be completed, or else a gap will remain in the critical philosophy.
>
> —Kant, letter to Christian Garve from September 21st, 1798 (*Correspondence* 12:257)

0.0 THE POST-CRITICAL KANT

In the Preface to the *Critique of the Power of Judgment* (*CJ*), Kant claims to have brought his 'entire critical enterprise to an end.' Even if the Critical philosophy is itself complete, however, Kant holds that the 'doctrinal' task remains of presenting the 'metaphysics of nature and of morals.'[1] Although Kant arguably completes the latter task in 1797 with the publication of the *Metaphysics of Morals* (*MM*), Kant never published a corresponding *Metaphysics of Nature.*[2]

All we have is the *Metaphysical Foundations of Natural Science* (*MFNS*), a '*special* metaphysical natural science' which determines the empirical concept of matter (the movable in space) in accordance with the 'transcendental part' of the metaphysics of nature (*MFNS* 4:469–470).[3] In *MFNS,* Kant characterizes the transcendental part as that which deals with 'the laws that make possible the concept of a nature in general, even without relation to any determinate object of experience' (*MFNS* 4:469).[4] In the *Critique of Pure Reason* (*CPR*), Kant makes clear that these laws should be understood in terms of 'the understanding and reason itself in a system of all concepts and principles that are related to objects in general' (*CPR* A845/B873). In other words, the transcendental part of the metaphysics of nature is the Transcendental Analytic of *CPR* (the categories and their corresponding principles).[5] Kant's special metaphysical natural science concerns physics, the only natural science he considers possible in *MFNS.*[6] He believes, furthermore, that he has 'completely exhausted this metaphysical doctrine of body' (*MFNS* 4:473).[7]

Notwithstanding what he says in the Preface to *CJ,* at some point in the 1790s, Kant changes his mind with respect to the completeness of the Critical philosophy. In the above letter to Christian Garve, Kant claims that he feels a 'pain like that of Tantalus' given the problem of effecting a 'Transition [*Übergang*] from the metaphysical foundations of natural science to physics,' without which there will be a 'gap' in the Critical philosophy. Unfortunately, in his letter to Garve (and in a similar letter to Johann Kiesewetter one month later), Kant does not make clear exactly what the gap is or how the transition is supposed to bridge the gap.[8] Resolving these issues requires examining Kant's final unpublished work, the so-called *Opus postumum* (*OP*), which contains drafts of a new project entitled *Transition from the Metaphysical Foundations of Natural Science to Physics.*

According to an early biographer, L.E. Borowski, Kant considered *OP* to be the keystone [*Schlußstein*] of his philosophical efforts.[9] Even so, Kant scholars have largely ignored *OP.* Those who tackled it have reached little consensus on how it relates to the Critical philosophy and some believe that it represents a "post-Critical" period in Kant's philosophical development.[10] In some ways the classification "post-Critical" is helpful, though in other ways it is misleading. It is helpful insofar as it correctly suggests that Kant's post-Critical philosophy marks a major reorientation of how he views the Critical philosophy. At the same time, the label can be misleading insofar as it might suggest that the post-Critical philosophy constitutes a dichotomous break with the Critical philosophy that is as great as the perceived break between the Critical philosophy and the pre-Critical philosophy.[11] In sharp contrast, however, I will argue that the post-Critical period does not mark a dichotomous break with the Critical philosophy, but rather instead a radically important *elaboration* of it. In fact, I will argue that Kant's planned transition project is the post-Critical culmination of his Critical philosophy. This will become apparent, I believe, once one has identified the gap in Kant's Critical philosophy and seen how the transition project can fill the gap.

This introductory chapter aims to tell a story, both historical and philosophical, about Kant's final project that will explain the historical lack of scholarly work on *OP* as well as the lack of philosophical consensus among those who have worked on *OP*. When it comes to the latter, I will argue that these other interpretations face problems both textual and philosophical. I will offer an alternative interpretation of *OP*, one which aims to avoid the problems that these other interpretations face while inextricably linking the success of Kant's post-Critical transition project with the viability of the Critical philosophy. The remainder of the Introduction is broken into five sections. The next section will examine the structure and style of *OP* in order to both explain and debunk the view, held by some scholars, that Kant's final work can be dismissed as the product of a senile mind. Following that section, I will discuss the troubled history of the manuscript which offers a compelling reason for why Kant's final work has been historically neglected. After that section, I will describe the various phases of Kant's transition project in *OP* in chronological order. Next, I will present an overview of several major views on Kant's gap and the transition project in the secondary literature. Finally, the last section of the Introduction will present the structure of this book which aims to demonstrate the important role *OP* can play in completing Kant's Critical philosophy.

Put briefly, I believe that the gap in Kant's Critical philosophy lies in the transcendental part of the metaphysics of nature. Specifically, I will argue that Kant's theory of substance in the Analogies of Experience faces a dilemma he cannot overcome within the context of *CPR*. While I believe that Kant's transition project overcomes the dilemma and so bridges this gap, by doing so it also offers a solution to several other serious problems with the Critical philosophy, the most important of which is the problem of affection (i.e., is the subject affected in sensibility by things-in-themselves, appearances, or somehow both?).[12] The latter is a problem that attracted most early commentators to *OP* and has dogged Kant's Critical philosophy since its inception.

0.1 THE STRUCTURE AND STYLE OF THE *OPUS POSTUMUM*

The structure of the document within which Kant's transition project is found provides some reason for the lack of scholarly consensus concerning his project. Kant's reflections on this project are scattered throughout *OP*, which completely occupies volumes 21–22 of the *Akademie* edition of *Kants gesammelte Schriften*. Some additional leafs are published in volume 23 under '*Ergänzungen zum Opus Postumum*.' The main body of *OP* is broken up into thirteen *Convoluts* or fascicles. The fascicles were numbered by someone other than Kant, and although the numbers imply nothing about Kant's intended ordering of the manuscript, they are often used for reference. The ordering problem is partially mitigated by the fact that Erich Adickes spent four weeks in the summer of 1916 carefully determining the chronological order of the manuscript. Unfortunately, this order is not retained in the *Akademie* edition, although one can find a guide to Adickes' chronology at the end of volume 22.[13] But even if the pages are put in their correct chronological order, *OP* is less a completed philosophical work than it is a philosophical work-in-progress. Moreover, it contains not only Kant's drafts of a proposed major work in theoretical/natural philosophy entitled *Transition from the Metaphysical Foundations of Natural Science to Physics,* but also remarks on a number of other issues, including reflections on the smallpox epidemic, theological reflections on the nature of God, and moral reflections on the nature of practical reason.

Although these other reflections are interesting in their own right, this book will focus on Kant's transition project, which plays a central role through much of the manuscript. In order to keep my focus narrow, I will also ignore the impact that early German Idealism was having on Kant as he worked on the *OP*. Although it is clear that Kant was keeping abreast of philosophical developments in Germany and that he responded to some of these developments either implicitly or explicitly in *OP*, my primary concern is not with Kant's relationship to his contemporaries, but rather with Kant's relationship to himself.[14] Specifically, I want to identify the gap in the Critical philosophy and how the transition project helps to fill that gap while also solving certain other perennial problems with the Critical philosophy.

Even after narrowing one's focus, however, one's task is still quite difficult. Kant's writing style is paradoxical insofar as arguments are often at odds with one another when they are not repeated *ad nauseam*. This led some early scholars to dismiss *OP* as nothing more than the product of Kant's senility.[15] Other scholars, however, have provided compelling reasons for why this could not possibly have been the case.[16] There is good evidence that Kant's plan for a transition project dates back to at least 1790 in light of a letter that Kiesewetter wrote to Kant on June 8th, 1795. Here Kiesewetter inquires as to progress of 'the transition from your metaphysical foundations of natural science to physics,' a work Kant had promised 'for some years now' (*AK* 12:23.28–30).[17] Since Kant had broken off correspondence with Kiesewetter in 1790 and did not resume this correspondence until 1793 (during which there is no mention of the transition until 1795), the phrase 'for some years now' suggests that the plan for a transition dates back at least to 1790.[18] Consequently, Kant's plan for a transition project dates back to a time when he could not possibly have been senile (i.e., the same year that he published *CJ*). Kant continued to make important philosophical contributions after 1790, including works like *Religion within the Limits of Reason Alone* (1793), *Toward Perpetual Peace* (1795), and *MM* (1797). Kant was in the midst of developing the transition project when *MM* appeared and even mentions it in the latter.[19] Kant's mental powers did not noticeably decline until sometime in 1801, by which point the transition project was arguably complete.[20] In 1801, Kant was working on *Convolut* 1, the final fascicle. His writing is uncertain and he no longer seems capable of bringing a line of thought to its conclusion. His efforts will repeatedly begin just to stop and begin anew. The result is some things being repeated an inordinate amount of times, even by the standards of *OP*. For example, Kant formulates the title of his new work over 60 times and offers over 150 definitions of "transcendental philosophy." Finally, in a later portion of *Convolut* 1, Kant even seems to have lost his ability to number things in their proper order.[21] Notwithstanding the above evidence that Kant was sound of mind when working on the transition project, the most important thing, from an interpretive standpoint, is to evaluate Kant's mental state on the basis of the text rather than evaluating the text on the basis of Kant's purported mental state. Insofar as my reconstruction of Kant's project is coherent, compelling, and well-grounded in the text, it should serve as proof that Kant was sound of mind.

As I implied earlier, it is extremely important to judge *OP* not as a completed book, but rather as, at best, a working draft of something left uncompleted. Kant's repetitive writing style in *OP* has, furthermore, much in common with the way in which Kant drafted *CPR*. Both works were written on a number of individual sheets where each sheet contained a completed thought or argument. Often there were several drafts of the same argument, and Kant compared these versions against one another to determine which one was best suited to go forward. Here then is an illuminating analogy: To examine *OP* in order to determine the form of the transition project would be very much like examining Kant's notes from the 1770s in order to determine the form of *CPR*, if counterfactually we were not in a position to know how *CPR* actually came out. As Eckart Förster notes, in *OP*, much as in his other works, Kant follows his own advice about how best to write philosophy: 'First one writes down all thoughts as they come, without any order. Thereafter one begins to coordinate and then to subordinate' (*AK* 24:484).[22] The way in which some of Kant's formulations in *OP* seem to contradict one another reflect the first stage, whereas how some formulations repeat others reflects the second. Insofar as an internal argumentative structure develops, *OP* also reflects the third stage. In fact, some pages of *OP* are found in amanuensis copies, which was normally the fourth and final stage before publication.[23] Even so, it remains true that *OP* is unfinished, and any projection of its final form is at best reconstructive and speculative. As we will see below, this is even true of Kant's transition project since its systematic perspective changes over time.[24]

Insofar as the text of *OP* can substantiate conflicting interpretations, how should one distinguish good interpretations from bad ones? In what follows, I will follow four interpretive principles which, I believe, allow us to make this distinction:

1) A good interpretation should be maximally consistent with the text. This includes not only providing textual evidence for the interpretation,

but also addressing texts that seem to conflict with the interpretation. As I will argue, this requires not only taking into account *OP*, but Kant's Critical-era texts as well.

2) A good interpretation should make Kant maximally consistent with himself. Although one could make a textual case that Kant simply contradicts himself in *OP*, a good interpretation should, as best as possible, make sense of these apparently conflicting passages. Furthermore, since Kant believes the transition project fills an important gap *within* the Critical philosophy, a good interpretation should try to make the post-Critical Kant as consistent as possible with the Critical Kant. Otherwise, one risks filling the gap in the Critical philosophy only by sacrificing the Critical philosophy itself.
3) A good interpretation should be philosophically plausible. Interpretations that make Kant's view philosophically incoherent or fail to solve the philosophical problems they identify should be rejected in favor of interpretations that are philosophically coherent and fruitful.
4) A good interpretation should reflect Kant's intent for *OP*. As mentioned above, Kant considered *OP* the keystone of his philosophical efforts and he believed that the transition filled an *important* gap within the Critical philosophy. Holding all else equal, an interpretation that makes good on Kant's intent should be preferred to interpretations that are deflationary or hold that Kant misunderstood his own philosophical project.

An interpretation that follows all four principles should be preferred to an interpretation that fails to follow them. As I will argue in §0.4, all of the existing interpretations violate more than one of these principles. In the Conclusion, I will argue that my interpretation follows all of these principles and so should be preferred to other competing interpretations.

0.2 THE HISTORY OF THE MANUSCRIPT

Although rumors of senility and the unfinished nature of the manuscript have certainly not encouraged scholarly study over the years, scholarly neglect was also unavoidable given the manuscript's own troubled history.[25] Although *OP* was largely completed by 1801, it was not published in a form appropriate for philosophical study until 1936–1938, more than 130 years after Kant's death.[26] During this 130-year span, the manuscript itself fell into terrible disorder. It went through a series of private hands. People would borrow leaves from the manuscript and then return them with no concern for their original location, or in some cases not return them at all. It is thought that because of a great deal of dirt on one fascicle that the manuscript at one time fell to the ground and was put back together with no attention to its original order. And it is an actual fact that the manuscript was also itself lost for a period of about 50 years and then rediscovered in 1858 under a pile of books.

The first attempt at transcribing and publishing the manuscript took place under the direction of Rudolf Reicke in 1882 when several though not all parts were published in the *Altpreussische Monatsschrift.* There is no shortage of problems, however, with this version. Although Reicke provided a transcription of the text, he tasked his friend Emil Arnoldt with preparing the text for publication. Arnoldt subsequently deleted several parts of Reicke's transcription, changed punctuation and even entire sentences, without always noting his emendations.[27] Albrecht Krause purchased the manuscript of *OP* before the completion of Reicke's edition with the aim of publishing an unabridged version of the manuscript. Even though Krause died before seeing such a version completed, it remained his 'dying wish.'[28] Reicke's edition was so poorly suited for scholarly study that Adickes, in 1920, wrote a whole commentary on *OP* with the expressed purpose of convincing the *Preussische Akademie* to publish a new and unabridged version of *OP* as part of *Kants gesammelte Schriften.*

The manuscript did find its way in 1923 to de Gruyter, the publisher who was also putting out the *Akademie* edition of Kant's works. One might think this a happy outcome, but unfortunately the editor for de Gruyter, Artur Buchenau, and Adickes, the *Akademie* editor of Kant's *Nachlass,* could not agree on editorial principles. There was a deeper issue, however, that doomed Adickes' involvement in the project. According to Reinhard Brandt, Buchenau was concerned with Adickes' belief that Kant possessed a 'private metaphysics' which hinged on the objective reality of things-in-themselves. Buchenau simply did not trust Adickes to keep his interpretation of Kant separate from his editorial work on Kant's final manuscript.[29] The result of this infighting (and Adickes' subsequent resignation from the project) is that a number of passages are repeated both in the *Reflexionen* (with Adickes as editor) and *OP* (with Buchenau as editor). Of even greater concern to the reader, several passages are also inconsistently transcribed.[30] Most importantly, however, Adickes' dutifully reconstructed chronological ordering of *OP* was not retained in the *Akademie* edition. Buchenau (likely under the influence of his assistant Gerhard Lehmann) preferred to publish *OP* in the order he received the manuscript instead of attempting to reconstruct the order in which the manuscript was actually written.[31] Although this edition of *OP* occupies more than two complete volumes of *Kants gesammelte Schriften* and takes up 1,161 pages (the original manuscript has 527 written pages), it is obviously far from being a perfect edition of Kant's final work. In addition to the chronological and transcription problems with the *Akademie* edition, during the 130-year lull between Kant's death and final publication of *OP,* several leaves of the original manuscript were lost. Some were destroyed during World War Two. Some have been found since the *Akademie* printing.[32] Some are likely still out there waiting to be discovered. Although Buchenau might be forgiven for publishing those pages that were *unknown* to him, he also did not publish certain pages that were *known* to him though not in his immediate possession.[33] Most of these pages were printed in volume 23 of *Kants gesammelte Schriften,* though some have only been recently published.[34] Given the problems with the current version of *OP,* Adickes' heartfelt wish for a new version of *OP* bears repeating much as Krause's did before his. As Adickes says at the end of his commentary on *OP:*

> The *OP* is Kant's legacy for contemporaries and posterity, though uncompleted and imperfect. However, his intentions and major ideas step out with sufficient clarity from the darkness of the expressions in which they are often wrapped. And therefore it is an imperative obligation of respect, to let the whole work appear as soon as possible in a worthy form.[35]

The darkness that envelopes many of Kant's formulations in *OP* can also help to explain the difficulty that scholars have had in trying to reach consensus on the nature of the gap in Kant's Critical philosophy as well as its relation to the transition project. In many ways, *OP* can be viewed like a philosophical Rorschach test. It is easy for commentators to see what they want to see given the way *OP* is written and the overall state of the existing manuscript. Even so, I believe it is possible to adjudicate between interpretations on the basis of the principles I outlined at the end of the previous section. Before examining these interpretations, however, it is important to understand how Kant's transition project develops in *OP,* since this development makes it easier to comprehend the very different views commentators have adopted toward Kant's project.

0.3 THE PHASES OF KANT'S TRANSITION PROJECT

Once one has the chronological ordering of *OP* in hand, it is easier to understand the various phases of Kant's transition project. The first chronological page of *OP* contains a very critical anonymous review of *MFNS* from the *Göttingische Anzeigen* copied out in Kant's own hand.[36] The reviewer criticizes Kant's attempt, in the first proposition of the Dynamics section of *MFNS,* to ascribe a moving force to matter simply on the basis of phoronomical considerations, in this case, that 'nothing can abolish motion save motion in the opposite direction' (*OP* 21:415).[37] In the first proposition to the Dynamics, Kant defines matter as 'the *movable* insofar as it *fills* a space' (*MFNS* 4:496). Matter must resist motion into the space that it fills (resistance to penetration) through an opposite motion. Kant accounts for this opposite motion in terms of an original repulsive force that matter possesses. It is difficult to see, however, how a claim concerning moving force (original repulsion) could follow from a claim concerning mere phoronomy. According to the mechanical philosophy, a fundamental characteristic of any object is its motion or rest. This does not entail, however, that matter *must* be understood dynamically in terms of moving forces.[38] Matter can be in motion in accordance with natural law, without matter filling space through moving forces (one could be a corpuscularian).[39] Not surprisingly, the reviewer challenges Kant's inference from impenetrability to moving forces: '*Must* one think of a moving force in a wall, because, at the wall, one cannot progress further? It is not even clear how Phoronomy, which merely treats of motion without considering force (from which motion arises) could lead to moving force' (*OP* 21:415). It stands to reason that when Kant begins *OP*, he has begun to ask himself the same question.

Although the above is a challenge to Kant's general theory of matter in the Dynamics chapter of *MFNS*, commentators have also voiced concerns with his account of specific empirical phenomena in the General Remark on the Dynamics. In the latter, Kant ruminates on a number of issues in empirical physics, including the nature of density, the formation of bodies, and the differences between specific varieties of matter.[40] At the same time, however, Kant also makes quite clear that he is not trying to derive the specific determinations of matter *a priori* from his categorically determined concept of matter in general. As he says, 'But one should guard against going beyond that which makes possible the general concept of matter as such, and wishing to explain *a priori* its particular, or even specific, determination and variety' (*MFNS* 4:524).

When it comes to density at least, Kant was certainly aware of the problems that his account in the General Remark faced. In remarks written on a September 8th, 1792 letter from J.S. Beck as well as in his October 16th reply to Beck, Kant admits that his theory of density is circular though he regards the issue as a question for physics.[41] According to Kant, density depends upon the mutual limitation of original attractive and repulsive forces.[42] If there were only attractive forces, matter would reduce to a physical point, but if there were only repulsive forces, matter would disperse itself to infinity.[43] Attractive force is a penetrating force constant across all materials. Repulsive forces vary by material, but are only surface forces that account for impenetrability (solidity according to the mechanist) and that limit original attraction. The interplay of original attractive and repulsive forces determines 'the degree of the filling of a space with determinate content' (*MFNS* 4:525) or, put differently, the amount of matter within a given volume. According to Kant, however, 'the original attraction is proportional to the quantity of matter and extends to infinity' (*MFNS* 4:518).[44] Kant's theory of density is circular since original attraction is proportional to the density of a matter (or its quantity in a given volume), but the matter itself could not exist were it not for original attraction.[45] How can original attraction depend on the quantity of matter while the quantity of matter simultaneously depends on original attraction?[46]

In the early sections of *OP*, Kant returns to many of the issues that he first dealt with in the General Remark on the Dynamics. He seems intent on showing that *MFNS* can, in fact, be extended to empirical physics. At this early stage of *OP*, Kant's plan for a transition can be viewed as carrying out the doctrinal task he set for himself in the Preface to *CJ*, *viz.* the completion of his metaphysics of nature. Even if Kant's approach to empirical physics is very interesting in its own right, Adickes does not consider it fruitful. Although Kant's scientific discussions would be worthwhile if they proceeded from mathematical formulations and careful experiments, they are instead rife with inexact formulations and lofty speculations.[47] This leads Adickes to characterize the transition from *MFNS* to empirical physics as a '*fata morgana*.'[48] Put in terms of my third principle of interpretation (from §0.1), a characterization of Kant's transition project that relies primarily on these early sections of *OP* will be philosophically implausible.

Starting in 1798, Kant seems to have reversed his whole way of approaching the metaphysical foundations of natural science. Whereas in *MFNS* Kant takes an empirical concept of matter (movable in space) and then goes on to determine this concept in accordance with the categories, he now takes an *a priori* concept of matter (internally moving forces) and then goes on to determine this concept in accordance with the categories in order to generate an elementary system [*Elementarsystem*] of the moving forces of matter. Whereas *MFNS* presupposes an empirical concept of matter, *OP* (from 1798 forward) presupposes an *a priori* concept of matter.[49] Notwithstanding this epistemic difference, both have a special conceptual status within Kant's metaphysics of nature. Within *MFNS*, the empirical concept of matter is the unique conceptual basis for Kant's metaphysical doctrine of body.[50] Although many different things fall under the extension of the empirical concept of matter in *MFNS*, Kant holds that the object of his *a priori* concept in *OP* 'is also *unique* and not a concept which several things have in common' (*OP* 22:554).[51] Since Kant is concerned with an *a priori* concept in *OP*, he adopts the transcendental argumentative approach familiar from *CPR*. This is not surprising because Kant is now operating firmly within the transcendental part of the metaphysics of nature and is concerned with a transcendental principle rather than a (special) metaphysical principle.[52] Kant makes this distinction in *CJ*:

> A transcendental principle is one through which the universal *a priori* condition under which alone things can become objects of our cognition at all is represented. By contrast, a principle is called metaphysical if it represents the *a priori* condition under which alone objects whose concept must be given empirically can be further determined *a priori*.
>
> (*CJ* 5:181)[53]

Although Kant continues to discuss many of the problems left over from the General Remark on the Dynamics, he comes to recognize that any solution to these problems will be properly speaking transcendental, requiring the moving forces of matter—or what comes to be known as "ether"—making these phenomena possible in accordance with the categories. The ether does not have determinate physical properties (e.g., it cannot be weighed) but is rather that which makes these determinate physical properties themselves possible. Consequently, Kant's conception of the ether belongs to the transition from the metaphysical foundations of natural science to physics. Although I will discuss the ether at greater length in Chapter Two, it is important to note that it is not the static medium that Michelson and Morley failed to discover in the 19th century. Rather, it is an essentially dynamic metaphysical foundation of matter, a plenum of attractive and repulsive forces. Even though Kant has come a long way from the Dynamics chapter in *MFNS,* he remains fundamentally committed to the idea that matter itself is constituted of attractive and repulsive forces.

At the beginning of the *Übergang* section written between May and August of 1799, Kant's transition project takes on a much greater scope. It appears that Kant recognized that the dynamic ether will be required not only for the solution of certain problems in empirical physics (e.g., those dealt with in the General Remark to the Dynamics), but will also be required to safeguard the unity of experience. Specifically, the ether will be required to preclude the experience of empty space which Kant believes would violate the unity of the subject's spatial experience. Consequently, the ether will be a transcendental *material* condition for experience. Being transcendental, therefore, the deduction of this material condition must proceed *a priori.* Kant begins his deduction with the unity of experience and tries to prove the existence of the dynamic ether. As we will see in the next chapter, although Kant's Ether Deduction is anticipated by some of what he says in *CPR,* it nevertheless constitutes a dramatic rethinking of what transcendental arguments are capable of proving.[54] In this respect, it is fundamentally different from the kinds of transcendental arguments that he offers during the Critical period. Once the Ether Deduction is introduced in 1799, furthermore, its conclusion continues to play an important role throughout the remainder of the transition project.[55] As I will argue, the Ether Deduction holds the key to understanding both the gap in Kant's Critical philosophy as well as how the transition project aims to bridge the gap.

Kant wrote *Convoluts* 10–11 immediately after the Ether Deduction and finished them sometime in April of 1800. Whereas the *Übergang* section deduces the existence of the dynamic ether for the unity of experience, *Convoluts* 10–11 describe what changes must be made, both in number and empirical function, to the other transcendental *formal* conditions of experience (space, time, categories, and apperception), given the addition of this *material* condition of experience, in order to insure the unity of experience.

Immediately after finishing *Convoluts* 10–11, Kant began work on *Convolut* 7, which he finished sometime in December of 1800. This was a point where the reader can still be fairly certain of Kant's mental powers which, by most accounts, did not start to decline until sometime in 1801. *Convolut* 7 builds upon the results of the *Elementarsystem,* the *Übergang,* and *Convolut*

10–11. In *Convolut* 7 and especially in the *Selbstsetzungslehre* (doctrine of self-positing), which constitutes its main part, Kant revisits the problem of affection which plagued the Critical philosophy from its inception. Kant argues that affection involves not only the receptivity of the subject to the dynamic ether, but also the activity of the subject affecting itself. Both kinds of affection are required not only for the cognition of physical objects, but also for the self-positing of the embodied cognitive subject.

Kant's characterization of the transition project itself also seems to change as *Convolut* 7 progresses. Whereas up to this point the transition had been characterized as between the metaphysical foundations of natural science and physics, Kant now oscillates between this formulation and two others. The first describes the transition as between the metaphysical foundations of natural science and transcendental philosophy.[56] The second describes the transition as progressing from the metaphysical foundations of natural science to transcendental philosophy and then finally to physics.[57]

After *Convolut* 7 comes *Convolut* 1, which Kant began in December of 1800 and continued to work on until February 1803 (a year before his death in 1804). In *Convolut* 1, Kant seems to set aside further development of the transition project in favor of a larger work which will include the transition entitled: 'The Highest Standpoint of Transcendental Philosophy in the System of Ideas: *God*, the *World*, and *Man* in the World, Restricting Himself through the Laws of Duty' (*OP* 21:59). Not only does this work unite the ideas of God and the world in the practical and theoretical self-positing activity of the subject, but it is this new cosmology that comes to define Kant's final view of transcendental idealism. This has led some commentators to claim that *OP* is actually two works. Krause believes that the transition project was largely completed in *Convolut* 7 and that *Convolut* 1 is the draft of a new work.[58] In contrast, Adickes holds that *OP* consists of one work in two parts.[59] In either case, *Convolut* 1 differs significantly from the other fascicles both in topic and tone and I will largely ignore it when reconstructing Kant's transition project. *Convolut* 1 also shows distinct signs of Kant's decreasing mental powers and it is unclear to what extent we should view this final stage of *OP* as reflecting Kant's considered opinion on the nature of transcendental idealism.

In summary, one simple way of viewing the structure of *OP* is as a series of stages where each subsequent stage takes on a greater scope than the previous one. Whereas Kant begins by simply trying to explain the specific varieties of matter, he soon comes to believe that his explanatory principle (the dynamic ether) must also serve as a transcendental material condition of experience. This move from natural philosophy back to transcendental philosophy also leads Kant to reexamine the role of the subject in constituting her experience. Although his theory of *Selbstsetzung* is initially limited to the theoretical domain, he soon expands it to include the practical. This finally leads him to see the theoretically and practically self-positing subject as generating both the physical as well as the moral worlds. In the final analysis, the physical world governed by natural laws and the moral world governed by God's practical laws are united and made possible only through the subject and her spontaneous free-activity.

In what follows, I will focus on those phases of Kant's transition project which seem to bear the closest connections to one another. I will argue that this project begins in 1798 with the development of an *Elementarsystem* of the moving forces of matter and ends sometime in 1800 with Kant's reflections on affection and the transition project in *Convolut* 7. The following section offers an overview of the secondary literature that deals with Kant's transition project. Readers who are already familiar with the literature on *OP* should feel free to skip this section.

0.4 RECEPTION OF THE *OPUS POSTUMUM* IN THE SECONDARY LITERATURE

As mentioned above, there is great disagreement between scholars as to the nature of the gap in the Critical philosophy as well as on what role the transition project might play in filling this gap. Notwithstanding this disagreement, I believe there are four major strands of interpretation that correspond to different views on the location of the gap in Kant's Critical philosophy. (1) One strand of interpretation holds that the gap lies between the special metaphysical natural science that Kant articulates in *MFNS* and empirical physics. (2) Another strand holds that the gap is the result of a failure in *CJ* itself or in the relationship between *CJ* and Kant's other Critical works. (3) Yet another strand of interpretation holds that the gap lies in *MFNS* itself, which has ramifications for the completeness of Kant's view in *CPR*. (4) A final strand holds that the gap lies in *CPR* itself irrespective of the success or failure of Kant's other Critical works. I will ultimately defend a version of (4).

The first strand of interpretation offers perhaps the most straightforward account of both the gap in the Critical philosophy as well as the role of the transition project in filling that gap. Both Hansgeorg Hoppe and Dina Emundts hold that the gap in Kant's Critical philosophy simply is the gap between *MFNS* and empirical physics.[60] Effecting a transition between the two will bridge the gap. For Hoppe, this requires showing how the pure natural science articulated in *MFNS* can be extended to empirical physics. Hoppe holds that the key to effecting this transition lies in establishing the scientific status of empirical physics.[61] Establishing the scientific status of empirical physics requires showing that the empirical moving forces of nature which are subject to experiments can be ordered in a system and this will in turn insure that these experiments yield genuine empirical cognition.[62] According to Kant, without systematicity there is no guarantee that our experiments yield empirical cognition since there would be no way to connect and compare the results of these experiments with one another.[63] For Hoppe, the transition project requires showing that there are certain *a priori* concepts that mediate between *MFNS* on the one hand and experimental physics on the other, insuring that the latter is systematic.[64] For Hoppe, these mediating concepts [*Mittelbegriffe*] are the *a priori* concepts of the relations of moving forces that Kant develops in *OP*.[65] Since experimentation, according to Kant, involves the *inserting* of concepts, these intermediary concepts will be inserted into perception through physical experiments to insure that the results of these experiments count as genuine empirical cognitions.[66]

Whereas Hoppe sees a natural gap between *MFNS* and empirical physics, Emundts believes this gap is precipitated by weaknesses in Kant's discussion of empirical physics in the General Remark on the Dynamics in *MFNS*. The latter might be considered Kant's first attempt to bridge the gap between *MFNS* and empirical physics. As mentioned in the previous section, the General Remark deals with issues in empirical physics like the nature of density, the formation of bodies, and the differences between specific varieties of matter. According to Emundts, the General Remark runs into internal problems, most importantly the circularity in Kant's theory of density, and must be supplemented by the transition project in *OP*.[67] Indeed, according to Emundts, the transition replaces Kant's account in the General Remark on the Dynamics.[68]

Emundts believes the solution hinges on Kant's ether theory, a theory which seems to bear little connection to the theory of matter developed in the Dynamics of *MFNS*.[69] As a form of matter, however, it seems important that Kant explain the ether in terms of the foundational attractive and repulsive forces that constitute matter in the Dynamics.[70] Instead of doing this, Kant instead replaces the theory of matter in the Dynamics with a new theory of matter in *OP* based on the internally moving forces of the ether.[71] The ether consequently comes to take on a much greater role than just solving the circularity problem which led to the gap in the Critical philosophy. In the *Übergang* section, the ether serves as a 'subjective principle of experience,' similar in some ways to the principle of *purposiveness* in *CJ*, but one which denotes something *real*.[72] Like Emundts, I will argue that, for Kant, the ether is real and that its actuality is tied to the conditions of experience for creatures like us. I will develop this interpretation in Chapter Three.

Hoppe is extremely skeptical of Kant's claims in the *Übergang* section and poses a dilemma for Kant's view of the ether. Either (1) the unity of experience simply depends on the combinatory faculty of the understanding, in which case the assumption of the ether is trivial and so dispensable, or (2) one must assume a pre-established harmony between our understanding and the ether as a material principle of unity, but Kant rejects this kind of pre-established harmony in *CPR*.[73] These are worries that I will return to below in Chapter Four (§4.3).

When it comes to evaluating the plausibility of this first strand of interpretation, one must return to Kant's correspondence. Although Kant's letters to Garve and Kiesewetter suggest that the gap in the Critical philosophy and

Kant's plan for a 'transition from the metaphysical foundations of natural science to physics' are two sides of the same coin, Kant's earlier correspondence with Kiesewetter suggests that the plan for a transition dates back to at least 1790. If so, then the plan for a transition dates back to a time when Kant considered the Critical philosophy to be complete. At most, the original plan for a transition could only have to do with the doctrinal task of completing the metaphysics of nature, though the completion of this task would have no impact on the completion of the Critical philosophy itself. Since Hoppe and Emundts believe the transition is meant to fill the gap that lies between *MFNS* and empirical physics, its primary value would lie in the completion of Kant's doctrinal task.

Although this is enough to show that Hoppe has not identified a gap within Kant's Critical philosophy, could the shortcomings that Emundts identifies in the General Remark on the Dynamics constitute such a gap? If Kant viewed the General Remark as a continuation of his attempt, from the main part of *MFNS,* to carry out an *a priori* construction of matter, this would be a distinct possibility. As mentioned in the previous section, however, Kant considers the General Remark to be wholly speculative and not part of the special metaphysical natural science he develops in the rest of the book. Emundts argues, furthermore, that the circularity in Kant's theory of density is a problem only for the General Remark and not for the main part of *MFNS.*[74] For a failure in the General Remark to open up a gap in the Critical philosophy, Kant would have had to change his mind with respect to the proper scope of the Critical philosophy. Although Kant's original plan for a transition can be viewed profitably as replacing his discussion in the General Remark on the Dynamics, perhaps with the goal of completing his metaphysics of nature, Kant's desire to *apply* the Critical philosophy to the domain of empirical physics does not constitute a gap *within* the Critical philosophy itself.[75]

The first strand of interpretation violates my first principle of interpretation (from §0.1) insofar as it ignores much of what Kant had to say in *OP* post-1798, i.e., what Kant wrote *after* he sent his letters to Garve and Kiesewetter. The claim that Kant changed his mind with regard to the proper scope of the Critical philosophy also violates my second principle for interpretation insofar as it requires us to accept that the post-Critical Kant disagrees with the Critical Kant on the proper scope of the Critical project.[76] The first strand interprets Kant's transition project largely in terms of his speculations concerning empirical physics in the early leaves of *OP.* As mentioned in the previous section, these speculations are themselves implausible, which violates the third principle of interpretation. Finally, the first strand of interpretation is deflationary since the gap is no longer within the "Critical philosophy" understood in terms of the three *Critiques.* This violates my final principle of interpretation. It seems we must look elsewhere in order to discover the gap within Kant's Critical philosophy.[77]

The second strand of interpretation holds that the gap in Kant's Critical philosophy is a result of a failure in *CJ* itself or in the relationship between

CJ and Kant's other Critical works. Vittorio Mathieu represents the first view and Michael Friedman the second. Mathieu shares something in common with the first strand of interpretation insofar as he believes that the gap in Kant's Critical philosophy lies between *MFNS* and empirical physics and that a mediating concept [*Mittelbegriff*] is necessary to effect a transition between the two. *MFNS* is simply a system of concepts *a priori* which requires a mediating concept to transition from this science of nature in general to physics as a system of nature. Without this mediating concept, a gap opens up in Kant's Critical philosophy between the metaphysical foundations of physics and the concrete physics of our actual world.[78]

Mathieu holds that Kant made a first stab at this mediating concept in *CJ* with the inner purposiveness of nature, but this concept is only *subjectively* necessary. At best, one can treat physics only *as if* it were a unified teleological system of nature without any guarantee that it actually is so.[79] According to Mathieu, Kant recognized the limits of this approach and decided he needed to find another mediating concept, one that is not simply *subjectively* necessary but rather *objectively* necessary of nature. This leads to Kant's plan for a transition between *MFNS* and empirical physics. Although the gap lies between *MFNS* and empirical physics, it is the failure of *CJ* that produces this gap.

Mathieu believes that the concept of the ether is itself this intermediate concept and is a transcendental concept derived from the categories.[80] The object of this concept is, however, neither physical nor metaphysical, but is rather what Kant calls 'physiological.'[81] For Mathieu, this means the ether is a "*Mittelmaterie*" (mediate material) in much the same way as the concept of the ether is a "*Mittelbegriff*."[82] Given the categorical basis of the concept of the ether as a transcendental condition, Mathieu takes the ether itself to be only another transcendental *formal* condition of experience, similar in role to the categorically based Axioms of Intuition (quantity), the Anticipations of Perception (quality), or the Schematism (time-determination) of *CPR*. He claims that it is a 'schematism of the system of moving forces insofar as they can be thought *a priori*.'[83] Although Mathieu believes the ether unifies the subject's experience, since the ether is a mediate material, it can only be *thought* as underlying the unity of this experience.[84]

If the ether can only be thought and its physical existence cannot be established, it should lead one to wonder whether Kant has really advanced that far beyond the subjectively necessary assumption of purposiveness in *CJ*. If there is no advance, then Mathieu's view violates my third principle interpretation insofar as it fails to solve the philosophical problem that it identifies. Although I agree with Mathieu that the ether serves valuable formal and material functions in Kant's post-Critical theory, I will argue below that the ether exists not merely *intentionally* but rather *physically*. The ether is the actual *extension* of the *a priori* ether *concept*. Since Kant's original plan for a transition project can be traced back to at least 1790 and Mathieu does not recognize that Kant's view on the transition changed over time, his interpretation of the transition would require us to believe that Kant rejects his approach (in *CJ*) toward systematization as soon as (or even before!) he published it. Furthermore, if the transition project that Kant mentions in his 1798 letters to Garve and Kiesewetter is meant to correct a significant problem with *CJ*, why would Kant allow the third edition of *CJ* to be published in 1799 without addressing this problem? Not only is this chronology unlikely, but it violates my second principle of interpretation insofar as it makes Kant inconsistent with himself.[85] More importantly, although the idea of nature's purposiveness is assumed for the systematization of nature in *CJ*, this idea does not itself offer any way of actually systematizing nature *a priori*. Kant insists, in the first Introduction to *CJ*, that the idea of purposiveness 'does not ground any theory and does not, any more than logic, contain cognition of objects and their constitution' (*CJ* 20:204). Consequently, regardless of whether Kant became dissatisfied with the merely subjective necessity of purposiveness, this is irrelevant to his need for an *a priori* theory of systematization.[86] Mathieu's interpretation violates my first principle of interpretation insofar as it is inconsistent with the text. Finally, like the first strand of interpretation, Mathieu locates the gap as lying between *MFNS* and empirical physics. Filling this gap, however, would only be the completion of Kant's doctrinal task (as articulated in the Preface to *CJ*) and could not be considered a gap in the Critical philosophy itself. As mentioned above, this would violate my fourth principle of interpretation insofar as it is deflationary. For these reasons, Mathieu's interpretation must be rejected.

Whereas Mathieu thinks that the gap arises due to a shortcoming within *CJ*, Friedman believes the gap arises from a failure to coordinate the different *methods* of *MFNS* and *CJ*. Whereas *MFNS* relies on the constitutive function of *determining* judgement, which begins with transcendental principles and moves downward to the metaphysical principles of the doctrine of body through which these transcendental principles are applied to the empirical concept of matter, *CJ* relies on the regulative capacity of *reflecting* judgement, which begins with particular empirical facts and move upwards to these transcendental principles.[87] Friedman claims that this problem of methodological coordination reveals a gap in the Critical philosophy.[88] What guarantee is there that the findings of reflecting judgment will be consistent with the findings of determining judgment? Since Friedman does not draw a distinction between reflecting judgment in *CJ* and the regulative ideas of reason in *CPR*, Förster notes that the problem would go even deeper than Friedman intends for it to go. It would expose a gap between the Transcendental Analytic (determining judgment) and the Appendix to the Transcendental Dialectic (reflecting judgment/regulative ideas of reason) in *CPR*.[89]

One response to Friedman's worry would be to say that the results of determining judgment place a *constraint* on the speculative scope of reflecting judgment.[90] For example, if determining judgment establishes an *a priori* foundation for the law of universal gravitation in *MFNS* (a central function of the book according to Friedman), then any reflecting judgment inconsistent with this law would need to be rejected. The *certainty* of determining judgment must always hold precedence over the *speculation* of reflecting judgment. Within the context of 'particular natural laws,' Kant suggests reflecting judgment is limited to considering only 'that which is left undetermined' (*CJ* 5:180) by the universal laws prescribed by determining judgment. Friedman seems to admit as much at a certain point when he says that 'even the regulative use of reason . . . is itself ultimately grounded in the conditions of the possibility of experience.'[91] The conditions of possible experience as articulated in the Transcendental Analytic, however, can also be seen as the result of determining judgment. The Principles of Pure Understanding offer conditions of possible experience and are the result of determining judgment. Likewise, they also serve as the *a priori* foundation for the necessity of natural law.[92] For example, the Third Analogy provides an *a priori* foundation (causal community) for the necessity of Newton's Third Law (equality of action and reaction) which itself plays an important role in the argument for the law of universal gravitation. Friedman's interpretation violates my fourth principle of interpretation insofar as the gap that he identifies appears to be a philosophical pseudo-problem. Friedman's interpretation also faces textual worries insofar as Kant does not discuss the distinction between determining and reflecting judgment at any length in *OP*. If coordinating these two methods constituted the gap, one would think Kant would spend more time talking about them. More so than the first strand of interpretation, Friedman's interpretation of the gap requires the reader to almost completely ignore what Kant actually says in *OP*, which violates my first principle of interpretation.

Given the Kiesewetter correspondence, Friedman separates the gap problem from Kant's transition project. Friedman takes the transition to consist of Kant's attempt to establish the scientific status of chemistry, something Kant now thinks is possible given certain developments in chemistry that offer hope that it might qualify as a natural science.[93] More generally, Kant now thinks it is possible to work out *all* the specific varieties of matter *a priori* (including chemical varieties), a topic that he only allowed himself to speculate on in the General Remark to the Dynamics of *MFNS*. In this respect, Friedman's interpretation of the transition shares much in common with Emundts' view though it avoids the objections that face the latter since Friedman separates Kant's plan for a transition from the gap problem. As I will argue below, however, Friedman does not pay sufficient attention to how Kant's conception of the transition changes over time. This is in violation of my first principle of interpretation. When Kant finally recognizes a gap in the Critical philosophy, he redescribes the transition project so that he can fill this gap.

Unlike Mathieu, Friedman holds that Kant is asserting the physical existence of the ether in the Ether Deduction, but nonetheless believes that Kant's conception of the ether is ultimately incoherent. Friedman cites a number of passages to support his first point and then poses a dilemma to motivate his second point. Either Kant has a conception of the ether as light-ether (oscillating medium for the transmission of light) or caloric/heat-ether (where increase or decrease of caloric within a body explains its state of aggregation). The light-ether explains how we visually perceive bodies at a distance to us and can be tied to Kant's project in *MFNS* insofar as the light-ether puts us in contact with the celestial bodies from which Newton's argument for universal gravitation in Book III of the *Principia* begins (i.e., Newton's "phenomena"). At the same time, however, it cannot explain the specific varieties of matter which is the task of the transition project. Alternatively, the caloric can explain the specific varieties of matter, but has no connection to Kant's arguments in support of universal gravitation in *MFNS* and so could not be part of a transition from *MFNS* to empirical physics. Since these are the only two options in Friedman's opinion, the Ether Deduction fails.[94] This violates my third principle of interpretation insofar as Friedman's view of the ether is not philosophically plausible. As will become clear shortly, I believe that Kant's conception of the transition project in the Ether Deduction is very different from his original conception of the transition project. Insofar as Friedman is relying on Kant's original conception in his criticisms of the Ether Deduction, these criticisms are unfair and inconsistent with large sections of the text (contrary to my first principle of interpretation). More importantly, however, I will argue below that it is possible to view Kant's plenum of attractive and repulsive forces (the dynamic ether) as the *common cause* of both perceptual affection and body formation in a way that would avoid Friedman's dilemma.[95]

The third strand of interpretation holds that there are certain problems *internal* to *MFNS* that pose a problem for the completeness of the Critical philosophy as articulated in *CPR*. Although Burkhard Tuschling and Förster disagree as to the location of the gap, they both fall within this strand of interpretation. Tuschling believes that what Kant means by the 'metaphysical foundations of natural science' changes dramatically by the time he begins the transition project in *OP* and this change in meaning was precipitated by the failures of the Dynamics chapter of *MFNS*. As mentioned in the previous section, Kant's argument for original attractive and repulsive forces constituting matter in the Dynamics rests on a non-sequitur and his theory of density in the General Remark on the Dynamics is circular. These failures led Kant to introduce a new theory of matter in *OP*, *viz.* his ether theory. Consequently, when Kant mentions the 'metaphysical foundations of natural science' in *OP*, the only part of *MFNS* this might include is the first chapter on Phoronomy (pure doctrine of motion) since it is the only chapter that does not depend on the Dynamics.[96] For Tuschling, the transition project, which Kant first mentions in his letter to Kiesewetter in 1795, is simply an *ad hoc* way of justifying the development of his ether theory by connecting it with his project in *MFNS*.[97]

Although the failure of *MFNS* might seem limited to the Dynamics section, according to Tuschling, it actually reflects a weakness in the Critical philosophy as articulated in *CPR*. In *CPR*, Kant claims that *MFNS* is a 'test' [*Probe*] that the table of categories as presented in the Transcendental Analytic of *CPR* is 'indispensable in the theoretical part of philosophy for completely outlining the plan for a whole science insofar as it rests on *a priori* concepts' (*CPR* B109–110).[98] As mentioned at the beginning of this chapter, *MFNS* aims to determine the empirical concept of matter *a priori* through the categories. Given the problems that Kant's theory of matter faces in *MFNS*, it would seem that the latter fails the test for which it was designed. According to Tuschling, this failure has broad ramifications since Kant comes to view the ether (the core of his new theory of matter) as a condition for the possibility of experience generally and so the failure of his earlier theory of matter in the Dynamics of *MFNS* constitutes a gap in the 'critical theory of cognition *a priori*.'[99] Although I agree with Tuschling that the gap is in what Kant calls the 'transcendental part' of the metaphysics of nature and that Kant's new dynamical theory of matter in *OP* will play an important role in bridging the gap, I will argue that this gap does not depend on the success or failure of *MFNS*.

Returning to Kant's ether theory, in contrast to the above commentators, Tuschling argues that Kant does not conceive of the ether as a merely *hypothetical* material. The Ether Deduction aims instead to establish the *actuality* of a material *a priori*. Kant's argument for this material is intended to avoid the problems that plague the Dynamics section of *MFNS*, while the existence of this material must be assumed for the success of Kant's transition between the metaphysical foundations of natural science to physics. The transition requires showing that the ether is the ultimate object of physics. At the same time, this material is required for the possibility of experience.[100]

Tuschling is quick to point out, however, the paradox at the heart of Kant's transition project. Even if Kant is able to prove the existence of the ether, he is unable to prove that it is an object of physics on pain of undermining the very transcendental proof strategy of the Ether Deduction. If the ether were an object of physics, understood as an object of experience, it could *in principle* be cognized *a posteriori*. If the ether could be cognized *a posteriori*, however, it could not serve as the transcendental material condition of experience, since the transcendental conditions of experience must be ascertained *a priori*. According to Tuschling, this leaves Kant with no way to effect a transition from the metaphysical foundations of natural science to physics.[101] Tuschling's view violates my third principle of interpretation insofar as *OP* does not actually solve the philosophical problems it is supposed to solve. As I will argue in Chapter Four, Kant has a way of overcoming Tuschling's problems in *Convoluts* 10–11 of *OP*, but it requires that he modify the transcendental *formal* conditions of experience (both in number and empirical function) given the addition of the transcendental *material* condition of experience in the Ether Deduction. I will also argue that Kant broadens his conception of "physics" in the post-Critical period to match the ambitions of his transition project. Ultimately, Tuschling believes that

Kant assumes the transcendental reality of the ether as a condition for the possibility of experience which is incompatible with Kant's Critical-era transcendental idealism.[102] This violates my second principle of interpretation insofar as it sets Kant at odds with himself. As I will argue in Chapter Three (§3.4), Kant can be committed to the actuality of the ether without committing himself to transcendental realism.

Like Friedman, and for the same reason, Förster separates Kant's original plan for a transition project from his recognition of the gap in his Critical philosophy. Whereas Tuschling takes the Kiesewetter letter from 1795 as providing the first reliable evidence for a transition project, Förster argues convincingly that the phrase 'for some years now' requires us to think that Kant already had such a plan in mind when he published *CJ* in 1790.[103] If Förster is right, Kant's original plan for a transition and the gap problem are not two sides of the same coin as Tuschling (and other commentators) would have it.[104] Consequently, these commentators would violate my first principle of interpretation insofar as the plausibility of their interpretations requires ignoring conflicting texts.

What is the gap according to Förster, if it is separate from the original plan for a transition? For Förster, it has to do with finding a way of constructing the object of outer sense in general. Since the Schematism of *CPR* is limited to the time-determinations of inner sense, Förster believes *MFNS* must provide a schematism of outer sense in order to insure the objective validity of the categories.[105] Constructing the object of outer sense in general would offer 'examples (instances *in concreto*)' of the categories, consequently giving 'a mere form of thought sense and meaning' (*MFNS* 4:478).[106] The problem is that *MFNS* fails to construct the object of outer sense in general since Kant's theory of density is circular and so cannot provide any account of how bodies are formed without begging the question. With the discovery of a gap, Kant *redescribes* the transition as a 'schematism of the concepts of metaphysics' (*OP* 21:169).[107] Although Kant's original plan for a transition was separate from his recognition of the gap in his Critical philosophy, his discovery of the latter led him to redescribe the former so that he could bridge the gap.

Friedman doubts that the failure of *MFNS* could really undermine the objective validity of the categories. As mentioned above, Kant clearly distinguishes the "transcendental part" of the metaphysics of nature from its "special part" (i.e., *MFNS* as a special metaphysical natural science). If Förster is right, then the special part actually belongs to the transcendental part contrary to Kant's intention since it would be required to insure the objective validity of the categories, a task that was ostensibly completed in the Transcendental Analytic.[108] Kant also holds that the metaphysical doctrine of body (*MFNS*) should be dealt with in an 'isolated system' distinct from general metaphysics (*CPR*) and that doing so 'does not impair the completeness of general metaphysics' (*MFNS* 4:477). This suggests that any failure of *MFNS*, as an isolated system, would not have an impact on the completeness of the Critical philosophy as articulated in *CPR*. This is a serious textual problem for the third strand of interpretation which violates both my first principle of interpretation (textual consistency) as well as my second principle of interpretation insofar as it makes Kant inconsistent with himself.

In the Principles of Pure Understanding, Kant holds that 'in order to understand the possibility of things in accordance with the categories, and thus to establish the objective reality of the latter, we do not merely need intuitions, but always outer intuitions' (*CPR* B291).[109] Although Förster uses this passage from the B-edition of *CPR* to argue that Kant is relying on the spatial schematism provided by *MFNS* (published a year earlier) to apply the categories to outer intuition, Kant makes no mention of another spatial schematism being needed to fulfill this task and his arguments in the Principles rely simply on the temporally schematized categories as applied to outer sense.[110] It should also be noted that Kant never provides spatial schemata for the individual categories in *MFNS* as he does by way of time in *CPR* and even if one were to admit that the categorically derived concepts (or predicables) in *MFNS* are spatial schemata of the categories, it is difficult to determine what predicables schematize what categories (e.g., what categories do the concepts of motion, resistance, or force schematize)?[111] Finally, since Kant holds in the Transcendental Aesthetic of *CPR* that 'all representations, whether or not they have outer things as their object, nevertheless as determinations of the mind themselves belong to inner sense' (*CPR* A34/B50), one might wonder whether a separate schematism of outer sense is really necessary. Although for different reasons, Friedman's and Förster's interpretations violate my fourth principle of interpretation insofar as the gaps they identify seem to be philosophical pseudo-problems.

If Förster's interpretation is rejected, however, what 'excellent and indispensable service' (*MFNS* 4:478) does *MFNS* provide for *CPR?* Friedman believes that *MFNS* does provide an important service insofar as it is the 'very first application of the categories and principles of the *Critique* to an actual empirically perceptual domain' while demonstrating that this application must be made to instances in outer sense.[112] It is important to note, however, that even if *MFNS* fails in its task this would not undermine what Kant has already achieved in *CPR*.

Regardless of how *MFNS* might relate to the Critical philosophy, what should we make of Förster's criticisms of *MFNS* itself? I believe Förster might well be demanding too much of *MFNS*. Kant claims that the specific determinations of matter (including the formation of bodies and density) cannot be constructed *a priori*.[113] As mentioned above, although Kant speculates on these topics in the General Remark on the Dynamics, he does not consider these speculations to meet the standards of *a priori* construction. In fact, he explicitly warns against thinking that they do so. Furthermore, as Friedman notes, Kant never returns explicitly to the circularity issue in *OP*, which should give one pause in thinking that it is central to understanding the gap in the Critical philosophy.[114] Although again for different reasons, both Friedman's and Förster's interpretations violate my first principle of interpretation insofar as they seem to be inconsistent with the text itself. Notwithstanding these issues, it is still important to note that Kant's account both of matter in general in the Dynamics as well as the specific varieties of matter in the General Remark face important objections.

A final strand of interpretation holds that the gap in Kant's Critical philosophy can be traced back to *CPR* regardless of the success or failure of *MFNS*. As I will argue, the gap in Kant's Critical philosophy lies in the transcendental part of the metaphysics of nature that he presents in the Transcendental Analytic of *CPR*. In particular, I will claim that the gap stems from problems that face the Analogies of Experience in *CPR* and that Kant would face these problems regardless of whether *MFNS* is successful.[115] Although early commentators on *OP* were not very concerned with the transition project itself or identifying the gap that Kant acknowledges in the Critical philosophy, they did think that *OP* offered a possible solution to the problem of affection, a problem that has plagued *CPR* since its publication. In the next section, I will lay out the structure of my book, which not only aims to answer the questions with which more contemporary commentators are concerned (*viz.* the relationship between the transition project and the gap), but will also show how, in answering these questions, Kant will also have a response to the problem of affection.

0.5 STRUCTURE OF THIS BOOK

The rest of this book is organized into five chapters. The first chapter will examine Kant's theory of substance in the Analogies of Experience from *CPR*. I will argue that Kant requires two concepts of substance within the Analogies. One concept is of relatively enduring individual empirical objects, or what I call "substances." The other concept is of a sempiternal and omnipresent material, or what I call "Substance." Since Kant has only the *category* (*a priori*) concept of substance at his disposal in the Analogies, however, I will argue that his theory of substance faces a dilemma and that he lacks the tools, within the context of *CPR*, which would be necessary and sufficient to resolve this dilemma. Assuming that the three Analogies stand or fall together, if Kant is unable to resolve this dilemma, there will be a significant gap within the Critical philosophy. This gap will lie at the heart of Kant's project in the Transcendental Analytic and so in the transcendental part of the metaphysics of nature.

The second chapter will track the development of Kant's conception of the ether from *CPR* to *OP*, while also explaining how this development might serve to solve certain problems in the Critical philosophy. Whereas my primary focus in other chapters is on how *OP* solves certain problems from *CPR* (the most important of which is the gap problem), this chapter will identify some problems with Kant's other Critical-era works (specifically *MFNS* and *CJ*) that Kant's post-Critical philosophy will also aim to solve. Although the problems discussed in this chapter lie on the periphery of the gap I have identified in Kant's Critical philosophy, examining them helps to underscore the enormous role the ether plays in Kant's post-Critical philosophy. In addition, as we will see, the gap will need to be bridged in order to solve some of these problems.

The third chapter offers a formal reconstruction of Kant's Ether Deduction from the *Übergang* (transition) section of *OP*. As I mentioned above (§0.3), in the Ether Deduction, Kant attempts to prove *a priori* the existence of a dynamic force plenum or what he calls "ether." Kant recognizes that the transcendental *formal* conditions for experience that he presents in *CPR* (space, time, categories, and apperception) are not sufficient to guarantee the unity of experience. In addition to these cognitive conditions, the ether must exist as a transcendental *material* condition for experience. At the same time, Kant's *a priori* proof strategy in the Ether Deduction delivers an *a priori* concept of the ether that corresponds to Kant's concept of Substance in the Analogies and is importantly different from the category of substance. Although the Ether Deduction offers an *a priori* proof for the existence of Substance and provides Kant with the conceptual resources necessary to avoid the dilemma that faces his theory of substance in the Analogies, he must still explain how the *a priori* concept of the ether is applied for the unity of experience in order to finally bridge the gap in his Critical philosophy.

Chapter Four examines Kant's attempt to bridge this gap in *Convoluts* 10–11, which is intimately connected to his redescription of the transition from *MFNS* to physics. "Physics" is redefined as 'empirical cognition in a system' and the concept of the ether plays a vital role in physics since it serves as the 'formal principle of the unification of the moving forces in the subject for the unity of experience' (*OP* 22:339). In other words, physics consists in the unity of experience as realized in this system of forces. This is a far broader conception of "physics" than the one operative in *MFNS* or in the early leaves of *OP*. Just as physics is redescribed with Kant's redescription of the transition, so too are the metaphysical foundations of natural science themselves redescribed insofar as the transcendental part of the metaphysics of nature (presupposed by the metaphysical foundations) is modified through the addition of the ether.

Kant argues that the ether has not only a *material* role to play in the transition project, but ultimately a *formal* role as well. What Kant realizes is the transcendental formal conditions of experience must be revised in number through the addition of the *a priori* concept of the ether and in their empirical function given the addition of the ether as a transcendental material condition of experience. After explaining how the ether helps to safeguard the empirical reality of space and time, Kant discusses how the *a priori* concept of the ether is applied by apperception, through the Principles of Pure

Understanding (Axioms, Anticipations, Analogies, and Postulates), in order to generate the unity of experience. In order for human experience as a whole to be unified, there must also exist, corresponding to it, a unified causal-dynamic whole, i.e., the ether as a plenum of attractive and repulsive forces. This will complete the transition from a redescribed "metaphysical foundations of natural science" to his very broad conception of "physics." Far from being separate issues, Kant's transition project in *Convoluts* 10–11 and the gap problem are intimately connected and it is only by recognizing both the transcendental material and formal functions of the ether that the transition can be effected and the gap in Kant's Critical philosophy filled. Grasping both the formal and material roles of the ether also helps to make sense of Kant's seemingly conflicting claims on the ether's ontological status in *OP* which has fueled the ongoing disagreement on its status in the secondary literature.

The final chapter of this book will examine *Convolut* 7 of *OP* where Kant applies the results of his transition project to the problem of affection. As mentioned in the previous section, concerns about Kant's theory of affection have plagued the Critical philosophy since its inception and most early interest in *OP* centered on how it might be able to overcome these concerns. The basic problem can be stated as a dilemma. If the subject is affected by appearances, then the effects of affection (*viz.* these same appearances) would have to be their own causes. If the subject is affected by things in themselves, however, then the subject would have to apply the categories (e.g., causation) beyond the domain of their legitimate application (the bounds of sense).

Both Adickes and Norman Kemp Smith agree that Kant is offering a theory of "double affection" in *OP* in order to overcome the dilemma, i.e., things-in-themselves affect the noumenal self who then organizes the resulting representations into a world of appearances that affects the phenomenal self.[116] Kemp Smith believes double affection is Kant's way of defending both the realist and the phenomenalist aspects of his thinking. Although Adickes goes to great lengths to defend double affection, Kemp Smith argues that it cannot protect Kant from the charge of phenomenalism since it places too much emphasis on the creative capacities of the subject. As he says, 'The self can be a knower only if it is a creator. An untenable method of distinguishing appearance and reality, backed by a subjectivism of the most extreme type, are the foundations upon which Kant is attempting to erect a realist view of the natural world!'[117]

Neither Adickes nor Kemp Smith is overly concerned with the relationship between the gap Kant acknowledges in his Critical philosophy and his planned transition project. Skeptical of Kant's *a priori* speculations concerning empirical physics, as mentioned above (§0.3), Adickes refers to the transition project as a "*fata morgana*" and for the same reason Kemp Smith says that Kant's attempts at a transition were 'from the start doomed to failure.'[118] Since they dismiss Kant's original plan for a transition project out of hand (the plan that many contemporary commentators embrace!), they do not take into account how this project changes over time. In particular, they do not attend to the way in which Kant redescribes the transition project in order to fill the gap he discovers in his Critical philosophy. As I will argue in the final chapter, once one has a proper understanding of the gap as well as how the transition fills the gap, one can view the passages that Adickes and Kemp Smith cite in support of double affection as supporting an alternate view of affection that overcomes the dilemma Kant's theory of affection faces while remaining wholly consistent with his Critical philosophy post-Critically understood.

NOTES

1. For similar claims, see *CPR,* Axxi and Bxliii. In order to understand how *CJ* completes the Critical philosophy, it is important to understand how the three Critiques relate to one another. In brief, whereas *CPR* (and associated texts) articulates Kant's Critical-era theoretical philosophy and the *Critique of Practical Reason* (and associated texts) articulates Kant's Critical-era practical philosophy, *CJ* explains how these two seemingly independent aspects of Kant's philosophical thought might be coordinated with one another through the reflecting power of judgment.
2. In the Preface to *MM,* Kant claims that *MFNS* is the 'counterpart' of *MM.* See *MM* 6:205. This might suggest that Kant believes *MFNS* constitutes his metaphysics of nature just as *MM* constitutes his metaphysics of morals. If this were true, however, Kant would have already completed, in 1786, the task he set for himself (i.e., laying out a metaphysics of nature) in 1790. As Kant makes clear in a 1785 letter to Christian Schütz, *MFNS* only provides 'concrete examples' that will help to make his yet to be completed metaphysics of nature 'comprehensible' (*Correspondence* 10:406).
3. Kant claims that 'the basic determination of something that is to be an object of outer senses had to be motion, because only thereby can these senses be affected' (*MFNS* 4:476). This is an interesting change from *CPR* where motion is only an *accident* but not the *essential* property of matter (*CPR* A186/B229). In *MFNS,* Kant goes on to determine this empirical concept of matter in accordance with the categories of *CPR* and each chapter of *MFNS* corresponds to a different set of categories. See *MNFS* 4:477.
4. Michael Friedman identifies these "laws" with the Analogies themselves. See Michael Friedman, *Kant's Construction of Nature: A Reading of the Metaphysical Foundations of Natural Science* (Cambridge: Cambridge University Press, 2013), 25. In the next chapter, I argue that the gap in Kant's Critical philosophy lies in the Analogies and so in the transcendental part of the metaphysics of nature.
5. For a similar view, see Friedman, *Kant's Construction of Nature,* 564. Konstantin Pollok also includes the principles of the Transcendental Aesthetic in the transcendental part of the metaphysics of nature. Although I think this is correct, Kant's primary focus in *MFNS* seems to be the determination of the empirical concept of matter *a priori* in accordance with the principles of the Transcendental Analytic. See Pollok, *Kants Metaphysische Anfangsgründe der Naturwissenschaft: Ein Kritischer Kommentar* (Hamburg: Felix Meiner, 2001), 119. As we will see, however, modifications that Kant makes to the

transcendental part of the metaphysics of nature in *OP* lead him to reassess the Transcendental Aesthetic as well (see §4.1).

6. Kant's conditions for "science" are rather stringent. Its claims must be apodictically certain and ordered according to *a priori* rational principles. See *MFNS* 4:468–469. Kant rejects both chemistry as well as psychology as candidates for sciences. See *MFNS* 4:470–471. Friedman suggests that Kant had changed his mind with regard to chemistry by the time he started working on *OP.* See Friedman, *Kant and the Exact Sciences* (Cambridge, MA: Harvard University Press, 1992), 215–217, 238–242, and 287–290.
7. As we will see, Kant comes to rethink the metaphysical doctrine of body in *OP,* especially with regard to its dynamical foundation.
8. See *Correspondence* 12:258–259.
9. See L.E. Borowski, *Darstellung des Lebens und Charakters Immanuel Kants* in *Immanuel Kant. Sein Leben in Darstellungen von Zeitgenossen. Die Biographien von L.E. Borowski, R.B. Jachmann und E.A. Ch. Wasianski,* ed. Felix Groß (Darmstadt: Wissenschaftliche Buchgesellschaft, 1993), 183. Giovanni Pietro Basile notes that Kant often used "Schlußstein" in his published work and others that knew Kant in his final years (e.g., J.G. Hasse and R.B. Jachmann) describe Kant's estimation of his final work in a similar way. At the same time, E.A. Wasianski, Kant's executor, claimed that Kant went back and forth over the value of his final work. This led Wasianski to ask Kant's most trusted expositor of the Critical philosophy, Johann Schultz, to determine whether the manuscript was fit for publication (for Kant's estimation of Schultz, see *Correspondence* 12:367–368). Schultz's opinion was that the manuscript should not be published. Basile rightly notes that Wasianski should not have relied on only one opinion and argues that Schultz may not have wanted *OP,* a manuscript that seemingly breaks from the Critical philosophy, to be published. See Giovanni Pietro Basile, *Kants Opus postumum und seine Rezeption* (Berlin: De Gruyter, 2013), 450–458.
10. This phrase was coined by Eckart Förster, "Kant's Notion of Philosophy," *The Monist* 72, no. 2 (1989): 285.
11. By "pre-Critical," I am referring to Kant's writings prior to his publication of the A-edition of *CPR* in 1781. By "post-Critical," I am referring generally to Kant's writings after his publication of *CJ* in 1790, though in particular to *OP.* For Frederick Beiser, the central interpretative question of *OP* is whether it is consistent with his Critical-era transcendental idealism or if it indicates a turn toward post-Kantian absolute idealism. See Frederick Beiser, *German Idealism: The Struggle against Subjectivism, 1781–1801* (Cambridge, MA: Harvard University Press, 2008), 182. Although I think this dichotomy is too simple (e.g., some interpreters view *OP* as a return to pre-Critical transcendental realism), I do believe Kant's post-Critical writing is largely consistent with his Critical-era transcendental idealism. It should also be noted that not everyone thinks that Kant's Critical philosophy constitutes a dichotomous break with the pre-Critical philosophy. For example, see Eric Watkins, *Kant and the Metaphysics of Causality* (Cambridge: Cambridge University Press, 2005).
12. I will articulate this problem in greater detail at the end of this chapter.
13. There is a less detailed chronological chart in Förster's Introduction to his and Michael Rosen's English translation of *OP,* xxvii. For our purposes, the key phases, in chronological order, are the *Elementarsystem* (1798–1799), the *Übergang* section (1799), *Convoluts* 10–11 (1799–1800), and *Convolut* 7 (1800).
14. For a discussion of how the post-Critical Kant relates to his contemporaries, see Förster, "Fichte, Beck, and Schelling in Kant's opus postumum," in *Kant and His Influence,* eds. George McDonald Ross and Tony McWalter (Bristol: Thoemmes, 1990), 146–169. For several essays that deal with how the

post-Critical Kant relates to later developments in German Idealism, see Ernst-Otto Onnasch, ed., *Kants Philosophie der Natur. Ihre Entwicklung im Opus postumum und ihre Wirkung* (Berlin: De Gruyter, 2009).

15. A defender of the senility interpretation might also note how Kant complains about being 'mentally paralyzed' (*Correspondence* 12:257) and an 'invalid' (*Correspondence* 12:258) in his 1798 letters to Garve and Kiesewetter. When it comes to Kant's mental paralysis, this might have more to do with the difficulty of the transition project itself rather than any mental weakness on Kant's part. When it comes to being an 'invalid,' Kant had long complained about his age and weakness going back to at least 1785 (when he was in the midst of working on the Critical philosophy!). See *Correspondence* 10:406–407. For a good discussion of these issues, see Basile, *Kants Opus postumum,* 440–444.
16. While commentators like Kuno Fischer and Benno Erdmann thought that *OP* could be dismissed as a product of senility, commentators like Erich Adickes and Förster have devoted considerable effort toward (in my opinion) very successfully debunking this idea. For more on this issue, see Part One of Erich Adickes, *Kants Opus postumum: dargestellt and beurteilt* (Berlin: Reuther & Reichard, 1920), as well as Förster's Introduction to his translation of *OP*.
17. Translation is mine.
18. Förster draws this conclusion from the Kiesewetter correspondence. See the Introduction to his translation of *OP*, xxviii.
19. See *MM* 6:468.
20. In §0.3, I will provide some reason for separating the transition project from *Convolut* 1.
21. See Adickes, *Kants Opus postumum,* 148–154.
22. Quoted by Förster in the Introduction to his translation of *OP*, xxiv.
23. I will return to the issue of amanuensis copies again when discussing Kant's Ether Deduction in Chapter Three (§3.1).
24. Reinhard Brandt makes this point. See Reinhard Brandt, "Kants Vorarbeiten zum *Übergang von der Metaphysik der Natur zur Physik* Probleme der Edition," in *Übergang: Untersuchungen zum Spätwerk Immanuel Kants,* eds. Siegfried Blasche and others (Frankfurt: Klostermann, 1991), 16.
25. For more on the history of the manuscript, in English, one may refer to Förster's Introduction to his translation of *OP*, xv–xlvii. The most complete account, however, is in German: Basile, *Kants Opus postumum,* appendix two.
26. Although Adickes dates the beginning of *OP* to 1796, Förster dates it sometime between 1788 and 1790 given the correspondence between Kant and Kiesewetter.
27. Basile notes that, in correspondence, Arnoldt expressed his desire for Reicke to publish the manuscript with as few of his changes as possible. See Basile, *Kants Opus postumum,* 467. See also Emile Arnoldt, *Gesammelte Schriften* (Berlin: Cassirer, 1911), part two, vol. 4, 372–373.
28. See Adickes, *Kants Opus postumum,* 26.
29. See Brandt, "Kants Vorarbeiten zum *Übergang,*" 11.
30. For example, compare *AK* 18:679.1–9 (*Reflexion* 6352a) with *OP* 21:337.23–338.05.
31. Gerhard Lehmann's resistance to publishing the manuscript thematically and chronologically as Adickes suggested is based upon his own "private philosophy" whereby the whole of something must be given prior to any subsequent decomposition. Lehmann saw Adickes' decomposition of the manuscript as arbitrary, but as Brandt notes, publishing the work in the accidental order in which it was found was far more arbitrary. See Brandt, "Kants Vorarbeiten zum *Übergang,*" 13–14.

32. For the most recent example, see Förster, "Zwei neu aufgefundene Lose Blätter zum *Opus postumum,*" *Kant-Studien* 95, (2004): 21–28.
33. For a list of pages not included in vols. 21–22, see Brandt, "Kants Vorarbeiten zum *Übergang,*" 23–24.
34. See Werner Stark, "Loses Blatt Leipzig 1. Transkription and Bemerkungen," in *Übergang: Untersuchungen zum Spätwerk Immanuel Kants,* 146–156.
35. Adickes, *Kants Opus postumum,* 855. For what a more "worthy form" might look like, see the (as of yet, incomplete) new online version of *OP* : http://telota.bbaw.de/kant_op/. This version includes scanned images of the original handwritten pages as well as raw transcriptions of a few of them.
36. The reviewer was Abraham Kästner. See Oscar Fambach, *Die Mitarbeiter der Göttingischen Gelehrten Anzeigen, 1769–1836* (Tübingen: Universitätsbibliothek, 1976), 134. Cited by Förster, *Kant's Final Synthesis: An Essay on the Opus postumum* (Cambridge, MA: Harvard University Press, 2000), 183n17.
37. For the phoronomical proposition, see *MFNS* 4:490.
38. The fundamental difference between mechanism and dynamism is that the former conceives of matter as irreducibly given with forces as its properties whereas the latter reduces matter into relationships of original forces. For a good summary of the differences between the two views, see Basile, *Kants Opus postumum,* 397.
39. For Friedman, the crux of Kant's argument in the first proposition of the Dynamics is that, in the case of the dynamical resistance of one matter to penetration by another, velocity at the turn-around point is always well-defined. This is in contrast to the mechanist view which holds that this resistance is absolute in certain cases (e.g., a single corpuscle or a composite corpuscular body without vacant internal vertices). In such cases of perfect uniform reflection, velocity would not be well-defined at the turn-around point. See Friedman, *Kant's Construction of Nature,* 114 and 231–232. If determining the velocity at the turn-around point, within a dynamical framework, requires a law of repulsive force, however, it would seem that Kant's approach is in trouble insofar as he has difficulty establishing such a law in *MFNS.*
40. See *MFNS* 4:523–535. Although I will present Kant's claims in the General Remark as speculations, Friedman suggests that the General Remark has a deeper purpose. He argues that Kant's aim in the General Remark is to resolve the empirical issues under consideration through a progressive and open-ended experimental research program, guided by the regulative use of reason, and following his preferred continuum model of matter that he develops in the main part of *MFNS.* See Friedman, *Kant's Construction of Nature,* 273. Even if Kant offers the methodology for an experimental research program in the General Remark, however, it is not a program that Kant himself pursues either in the Critical or post-Critical periods.
41. See *Correspondence,* 11:376. For the complete letters and Kant's notes, see *Correspondence,* 11:359–366 and 375–377.
42. See *MFNS* 4:518, 521, and 525.
43. See *MFNS* 4:508 and 511. Pollok notes that if matter were dispersed to infinity, it could not affect our senses since it would have no intensive magnitude. The *quality* of matter is essential, i.e., it must have intensive magnitude or a degree of influence upon sense. See Pollok, *Kants Metaphysische Anfangsgründe,* 273 and 343. For Kant, matter dispersed to infinity is not matter at all (*MFNS* 4:508).
44. See also 4:512, 514, 516, and 564. Although Kant views '*original attraction,* as the cause of universal gravitation' at one point (*MFNS* 4:541), the other passages together make a fairly strong case that Kant believes original attraction

is not merely the cause of universal gravitation but rather identical to it. Kant's comments in his letter to Beck also suggest that he identifies original attraction with Newtonian universal gravitation. See *Correspondence* 11:376. For a more detailed explanation of how gravitation and original attraction are identical, see Martin Carrier, "Kraft und Wirklichkeit: Kants späte Theorie der Materie," in *Übergang: Untersuchungen zum Spätwerk Immanuel Kants,* 212–215.

45. For further discussion of the circularity problem, see Förster, *Kant's Final Synthesis,* 33–45, Pollok, *Kants Metaphysische Anfangsgründe,* 307–316, Burkhard Tuschling, *Metaphysische und Transzendentale Dynamik in Kants Opus Postumum* (Berlin: De Gruyter, 1971), 46–56, and Kenneth Westphal, *Kant's Transcendental Proof of Realism* (Cambridge: Cambridge University Press, 2004), 190–197. The above should be considered only *one* way of formulating the circularity objection. As Emundts notes, there are several potential objections to Kant's theory of matter that could be formulated as problems of circularity. See Dina Emundts, *Kants Übergangskonzeption im Opus Postumum* (Berlin: De Gruyter, 2004), 77.
46. One suggestion would be to hold that density and attractive force are not discrete phenomena and can be defined in terms of one another. This suggestion ignores, however, the apparent causal connection between density and attractive force, i.e., that an increase in density is the cause of the increase in the degree of attraction. At the same time, however, original attraction is the partial cause of density.
47. See Adickes, *Kants Opus postumum,* 588–591. This provides additional reason for rejecting the idea, popular among some interpreters (see §0.4), that *OP* is supposed to somehow replace Kant's special metaphysical natural science in *MFNS.* The latter aims to construct mathematically the concepts that constitute the empirical concept of matter (e.g., the concepts of mass or quantity of matter). See *MFNS* 4:470 and Friedman, *Kant's Construction of Nature,* 529. These kinds of mathematical constructions, however, are largely lacking from *OP.*
48. Adickes, *Kants Opus postumum,* 162.
49. Classifying the concept of motion as "empirical" is itself a bit of a simplification. Although establishing the objective reality of the concept requires experience, the concept itself can be constructed *a priori* in pure intuition. See *CPR* B154–155. The objective reality of the ether concept, however, cannot be established in the same way as the empirical concept of matter in *MFNS.* Whereas (loco)motion is taken to be the basic determination of matter in *MFNS* since it is the only way the senses can be affected (*MFNS* 4:476), the ether is not itself locomotive (since it is omnipresent) and so cannot affect the senses in this way (*OP* 21:584). As I will argue in Chapter Three, the ether is a transcendental material condition of experience but not an object of experience.
50. See *MFNS* 4:472–473.
51. As we will see in Chapter Four, since the concept of the ether is both *unique* and *a priori,* there are concerns as to how it can be integrated into the conceptual framework of the Critical philosophy (*viz.* the table of categories). One should not be concerned, however, that the distinction between conceptual extension and intension is anachronistic. In his logic lectures, Kant distinguishes between the 'extension' [*Umfang*] of a concept and its 'content' [*Inhalt*] (or intension). For Kant, the content of a concept is inversely proportional to its extension. Consequently, it should not be at all surprising that the intension of the ether concept is quite rich indeed since its extension is singular. See *AK* 24:911–912 and 9:95.
52. Tuschling goes so far as to claim that when Kant refers to the 'metaphysical foundations of natural science' in *OP,* he is not referring to the 1786 work of that same name. He believes that this allows one to overcome apparent textual contradictions, e.g., defining matter as the movable in space (*MFNS*)

and defining matter in terms of internally moving forces (*OP*). See Tuschling, *Metaphysische und Transzendentale Dynamik,* 12.

53. For a similar discussion of this passage, see Friedman, *Kant's Construction of Nature,* 336–337.
54. In particular, I will argue that the Ether Deduction is anticipated by Kant's arguments in the Analogies. At the same time, however, the conclusion of the Ether Deduction is meant to be *constitutive,* which marks a dramatic change from Kant's view in the Transcendental Ideal, where it would seem that such a conclusion could, at best, be *regulative.*
55. On this point, I am following Jeffrey Edwards, "Spinozism, Freedom, and Transcendental Dynamics in Kant's Final System of Transcendental Idealism," in *The Reception of Kant's Critical Philosophy: Fichte, Schelling, and Hegel,* ed. Sally Sedgwick (Cambridge: Cambridge University Press, 2000), 64.
56. For example, see *OP* 22:129.
57. For example, see *OP* 22:86.
58. See Albrecht Krause, *Das Nachgelassene Werk Immanuel Kant's. Vom Uebergane von den metaphysicischen Anfangsgründe der Naturwissenschaft zur Physik, mit Belegen populär-wissenschaftlich dargestellt* (Frankfurt: Schauenburg, 1888), 93 and 126.
59. See Adickes, *Kants Opus postumum,* 600–604 and 732–737. Adickes claims that both *Convolut* 1 and *Convolut* 7 belong to the second part of Kant's project. Although I acknowledge that Kant begins to re-envision his project as *Convolut* 7 progresses, it also serves to complete the transition project that Kant began in the *Elementarsystem.*
60. Although Kant often mentions a "chasm" [*Kluft*] between the metaphysical foundations of natural science and physics (e.g., see *OP* 22:279), I will argue that this is not the "gap" [*Lücke*] Kant mentions in his letters to Garve and Kiesewetter. I will discuss the latter in greater detail in §1.3.
61. Hansgeorg Hoppe, *Kants Theorie der Physik: Eine Untersuchung über das Opus postumum von Kant* (Frankfurt: Klostermann, 1969), 69.
62. Hoppe, *Kants Theorie der Physik,* 72–81. Following Kant (e.g., *OP* 22:298), Hoppe identifies physics as an experimental science. See Hoppe, *Kants Theorie der Physik,* 87.
63. Hoppe, *Kants Theorie der Physik,* 74. See also *OP* 22:166.
64. Hoppe, *Kants Theorie der Physik,* 81–88. Kant uses the term '*Mittelbegriffe*' at several points in *OP.* For example, see 21:311.
65. Hoppe, *Kants Theorie der Physik,* 88–93. It is important to note that, for Hoppe, these are not concepts of the moving forces themselves, but only their relations. Hoppe rejects the idea that Kant is offering a material anticipation of experience. See Hoppe, *Kants Theorie der Physik,* 89 and *OP* 22:265. Just a couple of pages earlier, however, Kant suggests that we can anticipate the moving forces themselves *a priori.* See *OP* 22:263. Beiser also rejects this aspect of Hoppe's interpretation, holding that it is inconsistent both with the spirit and the letter of Kant's text. See Beiser, *German Idealism,* 635n10.
66. Hoppe traces this conception of experimentation back to the Preface of *CPR* (Bxii–xiv). See Hoppe, *Kants Theorie der Physik,* 114–118.
67. Emundts, *Kants Übergangskonzeption,* 15–26.
68. Emundts, *Kants Übergangskonzeption,* 150. Between August and September of 1798, Kant claims that *MFNS* is insufficient to effect the transition to physics since it can 'furnish no specifically determined, empirical properties' (*OP* 22:282). Karen Gloy adopts a position similar to Emundts' view, though she argues that the transition project develops (rather than replaces) the General Remark on the Dynamics. See Karen Gloy, *Die Kantische Theorie der*

Naturwissenschaft. Eine Strukturanalyse ihre Möglichkeiten, ihres Umfangs und ihrer Grenzen (Berlin: De Gruyter, 1976), 11.

69. For a sketch of Kant's solution, see Emundts, *Kants Übergangskonzeption,* 103–106. I will evaluate Kant's post-Critical solution to the circularity problem in §2.2.
70. See Emundts, *Kants Übergangskonzeption,* 113.
71. Emundts, *Kants Übergangskonzeption,* 153. Beth Lord believes that the transition project is at least partially precipitated by problems with the General Remark to the Dynamics. She holds that the transition project aims to explain how individual bodies arise from a unified system of forces. The solution lies in Kant's conception of the ether as an oscillating medium that accounts for the cohesion of bodies. Like other first strand interpretations, however, her account of the transition places too much emphasis on the early leaves of *OP* and does not recognize how Kant's view on the transition changed over time. See Beth Lord, *Kant and Spinozism: Transcendental Idealism and Immanence from Jacobi to Deleuze* (Basingstoke: Palgrave Macmillan, 2011), 157–158.
72. Emundts, *Kants Übergangskonzeption,* 200–201. See also 174 and 181–195. I will discuss Kant's views in *CJ* at greater length in connection with the second line of interpretation of *OP.*
73. Hoppe, *Kants Theorie der Physik,* 106. For Kant's rejection of pre-established harmony, see *CPR* B167–168.
74. Emundts, *Kants Übergangskonzeption,* 77–78.
75. Like Förster, Emundts recognizes that Kant already had a plan for the transition back in 1790. Emundts, *Kants Übergangskonzeption,* 25.
76. Although I will not challenge the idea that Kant changed his mind regarding, for example, the scientific status of chemistry, my interpretation of the transition project and the gap problem will not rely on Kant changing his mind.
77. For further discussion of Emundts' position, see Eric Watkins' review of her book in the *European Journal of Philosophy* 16, no. 2 (2008): 332–336.
78. Vittorio Mathieu, *Kants Opus postumum,* ed. Gerd Held (Frankfurt: Klostermann, 1989), 39–41. Although their interpretations differ, Mathieu acknowledges a debt to Lehmann who likewise connects Kant's transition project back to *CJ.* Even though they agree on the gap problem, Lehmann believes that Kant attempts to bridge this gap through reflective judgment and that *CJ* provides the form for Kant's discussion in *OP.* He claims that Kant's transition project in *OP* is analagous to the transition from the concepts of nature to the concepts of freedom that Kant mentions in the introduction to *CJ* (*CJ* 5:179). See Gerhard Lehmann, *Beiträge zur Geschichte und Interpretation der Philosophie Kants* (Berlin: De Gruyter, 1969), 188–194 and 295–299. Insofar as Lehmann's interpretation ascribes a major post-Critical role to reflective judgment, his view will face some of the same objections that Friedman's view faces below.
79. Mathieu, *Kants Opus postumum,* 42–44. See *CJ* 5:179–181.
80. Mathieu, *Kants Opus postumum,* 78 and 113. As I will argue in §4.2, even though there is a close connection between the categories and the concept of the ether, the latter is not simply derived from the former. Although Lord does not claim that the concept of the ether is derived from the categories, like Mathieu, she believes that it overcomes a problem left over from *CJ.* Although the concept of purposiveness can guarantee that natural laws are systematically ordered, it cannot guarantee an actual systematic unity of natural forces. Kant requires a transition from these natural laws to a systematic empirical doctrine by way of the ether concept which is an *a priori* principle of the unity of specific forces in nature. See Lord, *Kant and Spinozism,* 156–157. I will return to this issue in §4.4.

81. Mathieu, *Kants Opus postumum,* 97. See also *OP* 22:293 where Kant defines "physiological" as 'empirical representations of the aggregation of the moving forces of matter.' In the "Loses Blatt Leipzig 1," which can be dated to the 1790s, Kant holds that the 'physiological foundations of natural science' deal with 'matter insofar as it has moving forces.' See Stark, "Loses Blatts Leipzig 1," 146.
82. Mathieu, *Kants Opus postumum,* 119.
83. Mathieu, *Kants Opus postumum,* 138. As we will see below, Förster also notes the schematizing function of Kant's transition project, though it plays a much more central role in his interpretation.
84. Mathieu, *Kants Opus postumum,* 118 and 140. Förster seems to share Mathieu's view that the ether exists only in thought. Förster identifies the ether with a transcendental ideal since it is the idea of an individual thing that is determinable through the idea alone. See Förster, *Kant's Final Synthesis,* 91–92.
85. Förster makes this point. See Förster, *Kant's Final Synthesis,* 4.
86. Förster also makes this point. See Förster, *Kant's Final Synthesis,* 6–7.
87. For the distinction between determinate and reflective judgment, see *CJ* 5:179–181.
88. Friedman, *Kant and the Exact Sciences,* 254–256.
89. See Friedman, *Kant and the Exact Sciences,* 48, 251–253 and Förster, *Kant's Final Synthesis,* 189n24. Förster argues against Friedman's reading on both philosophical and textual grounds. See Förster, *Kant's Final Synthesis,* 7–11.
90. Förster makes a similar point. See Förster, *Kant's Final Synthesis,* 189n24.
91. Friedman, *Kant and the Exact Sciences,* 263.
92. For the relationship between the Principles of Pure Understanding and the laws of nature, see *CPR* A159/B198.
93. Friedman, *Kant and the Exact Sciences,* 215–217, 238–242, and 287–290.
94. Friedman, *Kant and the Exact Sciences,* 327–328 and 338–339.
95. When it comes to caloric, for example, Kant notes that heat 'may only be one particular effect of its moving forces' (*OP* 21:229).
96. Tuschling, *Metaphysische und Transzendentale Dynamik,* 88.
97. See Tuschling, *Metaphysische und Transzendentale Dynamik,* 31–32 and 61–65.
98. See Tuschling, *Metaphysische und Transzendentale Dynamik,* 37–39.
99. Tuschling, *Metaphysische und Transzendentale Dynamik,* 159.
100. See Tuschling, *Metaphysische und Transzendentale Dynamik,* 75, 100, 149, 176–178. For a similar approach to the ether's function see Jeffrey Edwards, *Substance, Force, and the Possibility of Knowledge. On Kant's Philosophy of Material Nature* (Berkeley: California University Press, 2000), 163–166.
101. See Tuschling, *Metaphysische und Transzendentale Dynamik,* 175–178.
102. See Tuschling, "Die Idee des transzendentalen idealismus in späten Opus postumum," in *Übergang: Untersuchungen zum Spätwerk Immanuel Kants,* 116.
103. See Tuschling, *Metaphysische und Transzendentale Dynamik,* 31 and Förster, *Kant's Final Synthesis,* 3.
104. It is important to note, however, that Tuschling does seem to have in mind a *redescribed* version of the transition project which is intimately connected with the gap problem. Unlike Hoppe and Emundts, he does not view the transition as dealing simply with an extension of *MFNS* to empirical physics, which would seem to be Kant's *original* plan for a transition. Beiser notes that even though Kant does not explicitly refer to the gap until 1798 he may still have discovered it earlier. See Beiser, *German Idealism,* 634n1. Basile provides some intriguing circumstantial evidence for the idea that the gap and the transition might both date from 1790. In an article delivered to Kant

immediately after the publication of *CJ*, Salomon Maimon claims that Kant requires a 'transition from general transcendental concepts and principles that refer to experience in general to the ones that refer to particular experiences.' Without this, Maimon claims that there are two gaps [*Lücken*] between: (1) the 'general transcendental and particular forms of things,' and (2) 'form and matter in general'. See Salomon Maimon, "Baco und Kant. Schreiben des H.S. Maimon an den Herausgeber dieses Journals," *Berlinisches Journal für Aufklärung* 7, no. 2 (1790):119–121. See also Basile, *Kants Opus postumum,* 376–380. Although Basile's evidence is highly suggestive, Kant never mentions Maimon (or this article) in his correspondence discussing the transition project/gap problematic or in the text of *OP* itself. Maimon's concerns, furthermore, are framed so generally that they are consistent with multiple strands of interpretation (including my own). Perhaps most importantly, even if Kant's *original* plan for a transition (from 1790) had to do with bridging the gaps that Maimon mentions, this does not preclude Kant from *redescribing* his transition project to fill a different gap that he discovers at a later point. Insofar as it is clear that Kant's plan for the transition project changes over time, it would suggest that his conception of the gap the transition is meant to bridge changes along with it.

105. For the limited function of the Schematism see Förster, *Kant's Final Synthesis,* 59.
106. See also *CPR* A240/B299 where Kant makes a similar claim.
107. See also Förster, *Kant's Final Synthesis,* 73.
108. Friedman, "Matter and Motion in the Metaphysical Foundations and in the First Critique," in *Kant and the Sciences,* ed. Eric Watkins (Oxford: Oxford University Press, 2001), 56. See also Friedman, *Kant's Construction of Nature,* 591. Pollok also believes that Förster's interpretation overlooks the division between the transcendental part (*CPR*) and the special part (*MFNS*) of the metaphysics of nature. When it comes to the question of *how* the categories relate to objects, Pollok further wonders whether *MFNS* really answers this question in a way that the Transcendental Deduction, Schematism, and Principles of *CPR* failed to do. Although Kant seems to downgrade the importance of the Transcendental Deduction in *MFNS* (4:475), it is unclear whether this downgrade is warranted. See Pollok, *Kants Metaphysische Anfangsgründe,* 3n7 and 142–149.
109. Emundts argues that Kant is not offering a *spatial* schematism of the categories, but simply applying the *temporally* schematized categories to space (outer sense). See Emundts, *Kants Übergangskonzeption,* 43–44. For his part, Westphal emphasizes the important role that outer intuition plays in applying the categories both in the A- and the B-editions of *CPR*. See Westphal, *Kant's Transcendental Proof of Realism,* 133–137.
110. For example, the schematized category of substance or 'persistence of the real in time' (*CPR* A144/B183) must apply to appearances in outer sense (space) since representations in inner sense (time) 'constantly flow' (*CPR* B291) and so cannot serve as the 'substratum of empirical time-determination in general' (*CPR* A144/B183).
111. Westphal makes this first point. See Westphal, *Kant's Transcendental Proof of Realism,* 133. Pollok makes a point similar to the second. See Pollok, *Kants Metaphysische Anfangsgründe,* 134–136. When it comes to the latter, Kant's own comments are not consistent. For instance, in *CPR,* he claims that force is a predicable of causality which is itself a category of relation (*CPR* A82/B108), but in *MFNS* he claims that concepts of force fall under the categories of quality (*MFNS* 4:523).

112. Friedman, *Kant's Construction of Nature,* 579. Although Friedman mentions the categories and the principles in the same breath, one could argue that the Principles of Pure Understanding offer many "instances *in concreto*" of the categories that give these forms of thought sense and meaning (e.g., the earth and moon in the Third Analogy illustrate causal community).
113. As mentioned above, both topics are dealt with in the General Remark on the Dynamics. As I also noted, Emundts argues that insofar as the problem with Kant's theory of density is internal to his account in the General Remark, it cannot be viewed as a problem that affects the main part of *MFNS* as Förster would have it. When it comes to the formation of bodies, one might wonder whether Kant's *physical* definition of a body (given in the General Remark) is required to accept his *mechanical* definition of a body (given in the main part of *MFNS*). See *MFNS,* 4:525 and 4:537. Even if Förster's concern was with the construction of the dynamical concept of *matter* rather than the construction of *body,* his view would still face problems. Although Kant deals with the possibility of mathematically constructing the dynamical concept of matter in remarks to Proposition Eight of the Dynamics, Friedman notes that Kant also expresses ambivalence about this project, separates it from his metaphysical treatment of matter, and only hesitantly presents his view. See Friedman, *Kant's Construction of Nature,* 223 and *MFNS* 4:522–523.
114. Friedman, *Kant and the Exact Sciences,* 223n8. Even if Kant does not return to the circularity problem explicitly in *OP,* his further remarks on cohesion and specific density are certainly relevant to its possible solution. See Westphal, "Kant's Dynamic Constructions," *Journal of Philosophical Research* 20, (1995): 409–414.
115. Insofar as I believe that the gap extends back to Kant's arguments in the Analogies, my view shares something in common with Westphal's view. See Westphal, *Kant's Transcendental Proof of Realism,* 146–166. At the same time, however, my concerns with the Analogies are not the same as Westphal's concerns. Even so, I will try to find a way of explaining how Kant can respond to Westphal's worries without committing himself to transcendental realism. I will discuss these issues at greater length below (see §1.1 and 3.4).
116. Hans Vaihinger was the first to develop this interpretation. See Vaihinger, *Kommentar zu Kants Kritik der reinen Vernunft* (Leipzig: Union Deutsche Verlagsgesellschaft, 1922), vol. 2, 51–53. Adickes and Norman Kemp Smith argue that *OP* provides confirmation of Vaihinger's view. See, Adickes, *Kants Opus postumum,* 18 and Norman Kemp Smith, *A Commentary to Kant's Critique of Pure Reason* (London: Macmillan, 1923), 613.
117. Kemp Smith, *A Commentary to Kant's Critique,* 618. Lehmann describes Kant's position in a similar fashion. See Lehmann, *Interpretation der Philosophie Kants,* 362.
118. Kemp Smith, *A Commentary to Kant's Critique,* 611.

1 A Dilemma for Kant's Theory of Substance

> All appearances contain that which persists (substance) as the object itself, and that which can change as its mere determination, i.e., a way in which the object exists.
>
> —Kant, Principle of the First Analogy in the A-edition (*CPR* A182)

> In all change of appearances substance persists, and its quantum is neither increased nor diminished in nature.
>
> —Kant, Principle of the First Analogy in the B-edition (*CPR* B224)

1.0 INTRODUCTION

The two formulations of the Principle of the First Analogy raise important questions as to how the term "substance" should be understood. The first formulation might suggest that relatively enduring empirical objects are substances since they persist through the alteration of their properties (e.g., a leaf changing its color). I will call these "substances" (plural lower case "s"). The second formulation, in contrast, seems stronger than the first and might suggest that there is only one sempiternal and omnipresent Substance whose quantum in nature is neither increased nor diminished. I will call this "Substance" (singular upper case "S").[1]

Since these are different and mutually irreducible conceptions of substance, I will argue that Kant faces a dilemma when applying the category (or *a priori* concept) of substance. Briefly stated, if the category of substance applies to Substance, then although this would ensure that experience takes place in a common spatiotemporal framework, one could not individuate substances and perceive their alterations. If the category of substance applies to substances, however, then although one could individuate these substances and perceive their alterations, the category would not pick out a common spatiotemporal framework for one's experience of substances. In neither case would there be a unified spatiotemporal experience of substances.

The first section of this chapter provides support, both textual and philosophical, for a Substance interpretation of the Analogies of Experience. The second section provides support for a substances interpretation. The third section explains the dilemma that Kant faces given his equivocal use of the term "substance" in the Analogies. I will argue that Kant's inability to resolve this dilemma using tools available to him in *CPR* reveals an important gap in the Critical philosophy, one that he aims to fill through the post-Critical transition project in *OP.*

1.1 THE SUBSTANCE INTERPRETATION OF THE ANALOGIES

The general principle of the Analogies is that 'experience is possible only through the representation of a necessary connection of perceptions' (*CPR* B218). In order to understand what this means, it is important to see how the Analogies fit into Kant's larger architectonic. The governing theory of *CPR* is transcendental idealism, which holds that we cannot cognize objects as they might exist in themselves but only insofar as they appear to us in accordance with the formal conditions of experience (space, time, categories, and apperception), where these formal conditions are contributions of the subject to her experience of these objects. All of these formal conditions, however, do not operate at the same cognitive level. Space and time operate at the level of sensible intuitions and are pure forms of intuition *a priori.* The categories are *a priori* pure concepts of the understanding which when combined with sensible intuitions produce cognitions. Apperception operates over the understanding and is the faculty by which unity is brought to the subject's cognitions *a priori* by applying the categories to sensible intuition through judgment. The three Analogies correspond to the categories of relation (substance/attribute, cause/effect, community) found in the Table of Categories, which in turn correspond to the relational forms of judgment (subject/predicate, hypothetical, disjunctive) found in the Logical Table of Judgments.[2] The relational categories are these judgments metaphysically interpreted as applying to an object of intuition in general rather than simply to propositions.[3] This object is not yet, however, an object of empirical or even pure intuition.[4] The Schematism will be required for the latter, whereas the Analogies will be required for the former.

The temporal schemata for the three categories of relation are persistence (substance), succession (cause and effect), and simultaneity (community).[5] In each case, Kant wants to say that, assuming transcendental idealism, the sensory objects given in experience will take on necessary temporal structures that are strictly transcendental, which is to say they are not derived from the objects but are rather imposed by the subject as a condition for the possibility of these objects. The result of the Analogies will be a unity of experience in apperception understood as a unity of appearances in time.

When it comes to substance in particular, the logical form of a subject/predicate proposition must first be metaphysically interpreted by applying it to objects in general. The result is the *a priori* concept or category of substance (an independently existing thing that supports properties), and its accidents (the contingent properties of the substance). Schematizing this concept requires giving it a temporal interpretation in pure intuition. The schematized category of substance is the notion of something which exists *persistently* [*beharrlich*] through time, and serves as the substrate for the various changes in properties that occur in time.[6] As a Principle of Pure Understanding, the First Analogy provides the rule for the objective use of the schematized category of substance in empirical intuition.[7]

Returning to the B-edition principle of the First Analogy, Kant believes that substance, which serves as the substrate of *all* change in appearance, is something that cannot change in its existence or its specific character, and whose quantum, consequently, can be neither increased nor diminished in nature. When denying that substance could "change" in its existence or its specific character, Kant uses the term "*wechseln.*" Henry Allison has argued that this term refers to replacement change in which one item is replaced by another. Although this definition of "*wechseln*" does not specify whether substances or states could *wechseln,* Kant's use of the term, in this context, suggests that only states or determinations of a substance can *wechseln,* but not the substance itself.[8] Elsewhere, Kant says that 'in all changes in the world substance remains and only the accidents change [*wechseln*]' (*CPR* A184/B227). Keeping in mind Allison's understanding of "*wechseln,*" one might say that since everyday objects of experience do arise and perish (e.g., Kant's example of how wood perishes through incineration while smoke and ash arise), these objects are mere alterations of Substance which does not itself arise or perish.[9] Allison argues that if there were no such persistent Substance, there would be no *common* framework or *single* backdrop by reference to which appearances could be determined to be either simultaneous with or successive to one another in a common time.[10]

Insofar as this backdrop is sempiternal, Allison's view suggests a Substance interpretation and Kant insists, in the First Analogy, that sempiternality is essential to his conception of substance. When discussing the successive nature of the apprehension of the manifold of appearance, Kant says that something must ground this manifold which '*always* exists, that is *remaining* and *persistent,* of which all change and simultaneity are nothing more than so many ways (*modi* of time) in which the persistent exists' (*CPR* A182/B226). Likewise, Kant insists that the 'the inner necessity of persisting is inseparably connected to the necessity of always having existed' (*CPR* A185/B229). Kant reinforces this conception of substance as something sempiternal at a later point of the First Analogy when he claims that 'everything that changes *remains* and only the *state* changes' (*CPR* A187/B230).[11]

Assuming that Kant thinks substance is sempiternal, why is the sempiternality of substance required for the unity of temporal experience? Even if this question can be answered, why should one think that substance must be omnipresent as well? Given that my conception of Substance is of something both sempiternal and omnipresent, both these questions must be answered. Finally, what relation does Substance bear to the other two Analogies of Experience? In what follows, I will attempt to establish (1) that substance must be sempiternal if the unity of temporal experience is to be maintained, i.e., if there is to be a common temporal framework for experience, (2) that the sempiternality of substance is necessary for the success of the Second Analogy, (3) that substance must be omnipresent if the unity of spatial experience is to be maintained, i.e., if there is to be a common spatial framework for experience, and (4) that the omnipresence of substance is necessary for the success of the Third Analogy. Bringing claims (1) and (3) together, I will argue that Substance (sempiternal and omnipresent) is necessary for there to be a unity of *spatiotemporal* experience, i.e., a common spatiotemporal framework for the subject's experience. Bringing claims (2) and (4) together, I will argue that Substance is necessary for the success of Kant's project in the other two Analogies as well. Beginning with (1), Kant claims:

> Alteration can therefore be perceived only in substances, and arising or perishing per se cannot be a possible perception unless it concerns merely a determination of that which persists, for it is this very thing that persists that makes possible the representation of the transition from one state into another, and from non-being into being, which can therefore be empirically cognized only as a changing determination of that which lasts.
>
> (*CPR* A188/B231)

Kant is claiming that whatever arises/perishes in time must be the alteration of something that persists, i.e., substance. The *absolute* arising of a substance requires that it arise from an empty time, otherwise it would arise from some prior existence and so the arising would not be absolute. Empty time, however, is no object of possible experience and offers no way of connecting this arising with anything previously existing within the temporal continuum. There is no reason to assume that an absolutely arising substance would arise within the *same* temporal continuum as everything else that exists in the temporal continuum since it arises from a time that does not include any of these things. Kant thinks the result would be *disconnected* temporal continuums which would violate the *unity* of time. To connect these temporal continuums with one another, one would have to assume a common time within which they overlap. Since time has only one dimension, however, does it even make sense to say that two times occupy the same time, i.e., exist simultaneously? As we will see in the Third Analogy, simultaneous existence in time only makes sense under the assumption of space. Even if we could make sense of two times existing simultaneously, what reason is there to assume that the two *times* could occupy the same common time if there is no reason to assume that the *substances* within these different temporal continuums occupy the same common time? Kant rejects the possibility of times *overlapping* one another side by side in favor of one time in which different times are placed *successively.*[12] As he says:

> Substances (in appearance) are the substrata of all time determination. The coming to be of some and the perishing of others would itself abolish the sole condition of the empirical unity of time, and the appearances would then be related to two different *times,* in which existence flowed side by side. For there is *only one time,* in which all different times must not be placed simultaneously, but rather one after another.
>
> (*CPR* A188–189/B231–232).[13]

Once one recognizes that substances cannot be experienced to either arise or perish absolutely, it is easy to see how Kant could claim that the quantum of substance is neither increased nor diminished in nature. An increase in the quantum of substance would require substance to arise absolutely (thus adding to the overall quantum) whereas diminishing the quantum of substance would require substance to perish absolutely (thus subtracting from the overall quantum).[14] If one accepts Kant's arguments in the First Analogy, then it seems as if substance must be *sempiternal* (cannot arise or perish absolutely) if there is to be a common temporal framework for the unity of experience.

When it comes to claim (2) above, *viz.* that the sempiternality of substance is necessary for the success of the Second Analogy, one should note that the conditions for the empirical unity of time presented in the First Analogy are presupposed in the Second Analogy. In the A-edition principle of the Second Analogy, Kant says that 'everything that happens (begins to be) presupposes something whereupon it follows according to a rule' (*CPR* A189). That any coming to be (i.e., arising) presupposes *something* (i.e., prior existence) rather than *nothing* (i.e., empty time) is required for the causal principle to function. Kant makes clear that if one were to suppose that nothing preceded an event, then there would be no way of applying a rule to this sequence such that it would be objectively determined rather than merely subjectively determined in apprehension. Without objective determination, however, one is left with 'only a play of representations relating to no object at all' (*CPR* A194/B239). Kant believes that an event following from an empty time can no more be apprehended than an empty time itself. As he says in the Second Analogy:

> That something happens, i.e. something or some state that did not exist before, cannot be empirically perceived where an appearance does not precede it that does not contain this state in itself. For a reality that follows upon an empty time, that is a coming to be preceded by no state of things, is as little capable of being apprehended as empty time itself.
>
> (*CPR* A191–192/B236–237)

Much as the sempiternality of substance is required for the empirical unity of time in the First Analogy, so too is it required for the objective determination of events in the Second Analogy. If substances could arise absolutely, then they would arise from empty times. Consequently, there would be events that would not follow from other events (i.e., prior existences) in accordance with a rule. This would, however, undermine the principle of the Second Analogy. Given this problem, it is not at all surprising that Kant held that *all* change of appearances is simply the alteration of some substance that persists.[15] It is important that Kant rule out the possibility of experiencing some change of appearances that is not the alteration of something that persists (*viz.* the experience of absolute arising or perishing) not only for the sake of his causal principle in the Second Analogy but for the empirical unity of time itself.

At this point, it is important to point out that what has been said so far does not establish a Substance interpretation of the backdrop thesis. One could understand substances not as relatively enduring but rather as sempiternal. There could be *many* (perhaps an infinite number) of sempiternal substances existing within the same temporal framework without the need for *one* sempiternal Substance to guarantee the unity of temporal experience.[16] Such substances seem sufficient for the empirical unity of time in the First Analogy as well as to safeguard the application of Kant's causal principle in the Second Analogy. When it comes to the former, any experience of arising or perishing is viewed simply as an alteration of a sempiternal substance that persists. When it comes to the latter, every event is viewed as an alteration of some sempiternal substance that persists where this event presupposes a previous event (prior state of a substance) from which it follows in accordance with a rule.

Notwithstanding the explanatory power of this interpretation, I believe that Substance is required for the unity of experience. This brings us to claim (3) above, *viz.* that substance must be omnipresent if the unity of spatial experience is to be maintained, i.e., if there is to be a common spatial framework for experience. Just as Kant argues that the *sempiternality* of substance is necessary to insure the unity of *temporal* experience by precluding the possibility of disconnected times, so too does he argue that the *omnipresence* of substance is necessary to insure the unity of *spatial* experience by precluding the possibility of disconnected spaces. In the Third Analogy, Kant holds that without 'matter everywhere' perceptions would be 'broken off' from one another (*CPR* A213–214/B260–261).[17] The existence of matter everywhere, however, precludes the possibility of experiencing empty space and consequently the possibility of perceptions being broken off from one another. The existence of matter everywhere insures the unity of the subject's spatial experience. In his Metaphysics Mrongovius lectures (1782–1783), Kant argues that substances separated by an empty time or space would occupy different worlds and so would be causally isolated from one another.[18] As mentioned above, a substance arising absolutely from an empty time would, in addition to violating the causal principle of the Second Analogy, occupy a different temporal continuum than any other substance. Such a substance would be causally isolated from these other substances. Likewise, substances separated by empty space could not causally interact and so could not stand in the causal community that the Third Analogy requires.[19]

This is not to say, however, that empty space, or empty time for that matter, is impossible, rather only the *experience* of empty space or time. Even if the omnipresence of substance is not necessary to preclude the experience of empty time, its omnipresence is necessary to preclude the experience of empty space. Its sempiternality and omnipresence, furthermore, are certainly sufficient to preclude both the experience of empty time and empty space while also insuring that substances do not arise absolutely (in violation of the Second Analogy) and can stand in causal community (as the Third Analogy requires).

The latter consideration leads directly to claim (4), *viz.* that the omnipresence of substance is necessary for the success of the Third Analogy. According to Kant, the reciprocal influence of matter everywhere is necessary for the causal community of substances and allows the subject to perceive their objective simultaneity. If there were absolutely empty spaces, the subject could not determine if appearances objectively follow one another (Second Analogy) or are simultaneous with one another (Third Analogy). Just as the sempiternality of Substance safeguards Kant's causal principle in the Second Analogy, so too does the omnipresence of Substance safeguard the distinction between objective succession and simultaneity in the Third Analogy. As Kant says when discussing two substances in space:

> For if you thought they were separated by a wholly empty space, so would the perception that proceeds from one to the other in time certainly determine the existence of the latter by means of a succeeding perception, however it could not be decided if the appearance follows the former objectively or is rather simultaneous with it.
>
> (*CPR* A212/B259)

Although I claim that both (3) and (4) are arguments for the omnipresence of substance, the connection between omnipresence and the category of substance may not be immediately obvious. In (3), it seems as if omnipresence follows not from the nature of substance, but rather from the requirements for the unity of spatial experience. Likewise, in (4) omnipresence follows not from the nature of substance, but rather from the requirements for distinguishing objective succession from simultaneity. Although commentators generally accept that substance must be sempiternal, few recognize its omnipresence.[20] Even though the Third Analogy seems to offer the best support for the omnipresence of substance, it is important to note that Kant does not mention 'substance everywhere,' opting instead for the locution 'matter everywhere.' Perhaps there is some distinction to be drawn between "matter" and "substance" within the context of the Analogies such that the two terms should not be substituted for one another. In the First Analogy, however, Kant does use the terms interchangeably. When giving the example of burning wood perishing as smoke and ash arise, he says:

> A philosopher was asked: How much does the smoke weigh? He replied: If you take away from the weight of the wood that was burnt the weight of the ashes that are left over, you will have the weight of the smoke. He thus assumed as incontrovertible that even in fire the matter (substance) never disappears but rather only suffers an alteration in its form.
>
> (*CPR* A185/B228)

Even if one grants that "matter" and "substance" can be substituted for one another within the context of the Analogies, what of the fact that omnipresence seems to bear no connection to the category of substance? Although it is easy enough to see how sempiternality can be connected to the persistence of substance, to what is the omnipresence of substance supposed to be connected? When it comes to the unschematized categories, it is important to note that neither temporal determinations nor spatial determinations are contained within their conceptual content. It is only through the schematization of the categories that they acquire their temporal determinations. Kant believes that these temporally schematized concepts are applied to outer sense (i.e., spatially) as a condition of their objective reality in the Principles of Pure Understanding.[21]

In the Axioms of Intuition (corresponding to the categories of quantity), Kant argues that all intuitions, whether they are pure or empirical, temporal or spatial, are going to have extensive magnitude. An extensive magnitude is one where the representations of its parts make possible the representation of the whole.[22] Kant uses the example of drawing a line in thought which involves 'successively generating all its parts from one point' (a spatial example) to illustrate the idea (*CPR* A152/B203). In the Anticipations of Perception (corresponding to the categories of quality), although the schematized category of reality is temporally determined as the 'quantity of something insofar as it fills a *time*' (*CPR* A143/B183), Kant applies this concept within the Anticipations to outer sense where he characterizes the real as filling both *space* and *time*.[23] Kant claims that a proof for empty space or time could not be based on experience since an entire absence of the real in sensibility could never be perceived nor could empty space or time be inferred from any instance of the real in sensibility regardless of its degree of reality.[24] Whether he is talking about extensive magnitudes or intensive magnitudes, Kant holds that what is given in outer sense is required to insure the objective reality of the schematized categories of magnitude. As he says, 'the possibility of things as magnitudes, and thus the objective reality of the category of magnitude, can also be exhibited only in outer intuition, and by means of that alone can it subsequently also be applied to outer sense'

(*CPR* B293). Just as in the Axioms and Anticipations, in the Analogies of Experience, Kant relies on outer sense to provide the schematized categories of relation with objective reality. When it comes to the schematized category of substance, Kant says:

> In order to give something that persists in intuition, corresponding to the concept of substance (and thereby to establish the objective reality of this concept), we need an intuition in space (of matter), since space alone persistently determines, while time, however, and thus everything that is in inner sense, constantly flows.
>
> (*CPR* B291)

Only what is given in outer sense (matter) persists and so is a condition for the objective reality of the schematized category of substance. While Kant's argument in the First Analogy shows that this matter must be sempiternal (existing at all times), one can hold that Kant's argument for "matter everywhere" in the Third Analogy shows that this matter must also be omnipresent (existing at all spaces). Whereas the unity of time requires that substance be sempiternal, so too does the unity of space require that substance be omnipresent.

Even if one accepts that there is an omnipresent sense of "substance" operative in the Analogies, it must be admitted that Kant himself equivocates between the singular and reference dividing uses of "substance," sometimes even in the same sentence.[25] Although sempiternal substances and sempiternal Substance are both arguably consistent with the B-edition principle of the First Analogy quoted at the outset of this chapter, as well as sufficient for safeguarding Kant's causal principle in the Second Analogy, Kant's use of the singular "substance" in the B-edition principle of the First Analogy and the role that "matter everywhere" plays in the Third Analogy seem to suggest a conception of sempiternal and omnipresent Substance rather than a conception of sempiternal and individuated substances.

Notwithstanding the textual support for a Substance interpretation, it is important to emphasize its philosophical functions as well. Substance is both *sempiternal* and *omnipresent*. Consequently, it insures the unity of *spatiotemporal* experience by precluding the possibility of experiencing empty times or spaces. It serves as the common framework within which the subject experiences objective succession (Second Analogy) as well as simultaneity (Third Analogy). The sempiternality of Substance safeguards the causal principle of the Second Analogy by insuring that everything that happens will in fact follow from *something* (rather than *nothing*), and the omnipresence of Substance is a condition for causal community of substances in the Third Analogy since it insures that these substances will occupy the same spatial framework. Finally, the omnipresence of Substance allows one to distinguish between the objective succession and simultaneity of substances. To put things in Allison's terms, an omnipresent (spatial) and sempiternal (temporal) backdrop (Substance) is required for the *spatiotemporal* unity of experience. Although Allison focuses exclusively on the sempiternality of this backdrop, I believe its omnipresence is equally important for the unity of experience.

One might object, at this point, that I am conflating empirical and transcendental conditions and that this has led me to offer a *metaphysical* interpretation of the Analogies (Substance *exists* as a necessary condition of the unity of experience) where an *epistemological* interpretation is more appropriate (a *concept* of Substance is a necessary condition of the unity of experience).[26] Although "matter everywhere" may be an *empirical* condition for a certain kind of experience in the Third Analogy (e.g., perceiving two substances in outer sense as existing simultaneously), the objector might claim that the category of community is the relevant *transcendental* condition. The crucial question, it seems to me, is whether "matter everywhere" is merely an empirical condition for a certain kind of experience or a transcendental condition of experience in general. As an *empirical* condition, it would have to be considered *contingent* relative to experience generally. I believe, however, that "matter everywhere" is a *necessary* condition for experience generally. As I will argue in §1.3, furthermore, Kant cannot establish the existence of "matter everywhere" empirically (*a posteriori*). Ultimately, what Kant will need is a transcendental (*a priori*) proof for the existence of Substance (or what he comes to call "ether").[27]

What lies behind the objection, I believe, is the assumption, common to the epistemological interpretation, that only *formal* conditions can be *transcendental* conditions. Admittedly, when Kant claims in *CPR* that 'the conditions of the *possibility of experience* in general are at the same time conditions of the *possibility of the objects of experience*' (*CPR* A158/B197), he has in mind the transcendental formal conditions of experience (space, time, categories, and apperception). As I will argue below, however, something material can serve as a condition for the possibility of experience in general without being an object of experience.[28] Elsewhere in *CPR*, Kant claims that 'there is only *one* experience in which all perceptions are represented as in thoroughgoing and lawlike connection' (*CPR* A110).[29] In other words, for Kant, experience is itself a unity (as we will see, this is an idea he returns to repeatedly in *OP*). Kant argues in the Third Analogy that "matter everywhere" is a condition for the thoroughgoing and law-like connection of perceptions and so would be a condition for *one* experience itself understood as the *unity* of experience. Consequently, "matter everywhere" seems to meet Kant's criterion for a transcendental condition of experience. More generally, that something omnipresent and sempiternal *exists* is a necessary condition for the unity of spatial (Third Analogy) and temporal (Second Analogy) experience. Something *must* exist that instantiates the concept of Substance (i.e., the concept must have extension) to insure the unity of spatiotemporal experience. As I will argue in later chapters, Substance is both a transcendental *formal* (conceptual) as well as transcendental *material* (ontological) condition for the unity of experience. Consequently, although my account of the Analogies might be more "metaphysical" than straightforwardly epistemological accounts, I in no way discount the value of the transcendental formal conditions as epistemological conditions of experience.[30]

Both Kenneth Westphal and I trace the gap in Kant's Critical philosophy back to the Analogies of Experience. For Westphal, the gap consists of the fact that Kant needs to establish the metaphysical causal thesis that 'every physical event has an external physical cause' which requires that there be spatially individuated substances.[31] Westphal does not believe that Kant establishes the metaphysical causal thesis since all three Analogies can be explained simply in terms of Substance.[32] In other words, Westphal takes the Substance interpretation one step further than I do by claiming it is not only necessary but also *sufficient* for explaining all three Analogies. Whereas the considerations of this section seem to support his reading when it comes to the First Analogy, he believes both the Second and Third Analogies can also be alternatively described using only Substance. With regard to the Second Analogy, the rule that every event has a cause could be accounted for simply by the living forces of Substance which brings about a time-ordered succession of its own alterations. With regard to the Third Analogy, the various substances that are supposed to stand in causal community could simply be modes of this same Substance.[33] Although I believe Kant needs Substance to make sense of the Analogies, I will argue below that the three Analogies cannot be adequately described using *only* Substance. Any adequate description of the Analogies will require both Substance as well as relatively enduring substances, and the next section attempts to establish the indispensability of the latter.

1.2 ANOTHER WAY OF UNDERSTANDING "SUBSTANCE"

Although I have provided support for the Substance interpretation, there is ample reason for thinking that ordinary empirical objects might function as substances in their own right. To begin, it is important to note that empirical intuitions are *immediately* related to their objects and are *singular.*[34] An appearance is the 'undetermined object of empirical intuition' (*CPR* A19/B33). Substance is the persistent *in* appearance. As Kant says, 'The persistent [*Beharrliche*], in relation to which all temporal relations of appearances can alone be determined, is the substance *in* appearance, i.e., the real *in* appearance' (*CPR* B225).[35] If only relatively enduring individual empirical objects appear, however, one might well wonder whether Substance itself could ever be the persistent in appearance. To put things in Allison's terms, one might wonder if the backdrop (or what I call "Substance") for the experiences of succession and simultaneity itself ever appears or if only successive and simultaneous substances placed against this backdrop appear. If the latter, and if the category of substance is the concept of Substance, then how is the category of substance ever applied to appearances? James Van Cleve raises a similar problem for Allison's backdrop thesis. Although we can *conceive* of the backdrop according to Van Cleve, we cannot *perceive* the backdrop.[36] If the category of substance is meant to apply to this backdrop, however, then to allow only its conceivability would be to admit the failure of the First Analogy insofar as it is a principle.[37] As mentioned above, the principles are meant to be rules for the objective use of the categories. If one is only able to conceive of Substance, however, then the category of substance would seem to have at best a *regulative* and not a *constitutive* use in the world of appearances.[38]

The nature of Substance itself also precludes it from being perceived. In *MFNS,* Kant claims that 'the basic determination of something that is to be an object of the outer senses had to be motion, because only thereby can these senses be affected' (*MFNS* 4:477). This is what leads him to define "matter" as the movable in space. Although Kant believes that the senses can be immediately affected only by pressure or impact, both of which require 'the approach [motion] of one matter to another' (*MFNS* 4:510), Friedman argues that Kant's point is more fundamental.[39] To understand this point, Friedman suggests examining two of Kant's other works: *Concerning the Ultimate Ground of the Differentiation of Directions in Space* (1768) and *What is Orientation in Thinking* (1786). In the former, Kant says, 'Concerning the things that exist outside ourselves: it is only in so far as they stand in relation to ourselves that we have any cognition of them by means of the senses at all' (*AK* 2:378).[40] Kant goes on to say that the subject must be embodied so that it can occupy a relative space which serves as the center of a three-dimensional framework. Keeping in mind that the concepts of relative space and relative motion are intertwined, the subject is affected only by objects that can be precisely located within this framework, i.e., objects that occupy discrete regions of space that move relative to the space that the subject occupies. In Chapter Five, I will refer to this as Kant's "causal-mechanical" theory of affection. Although substances (including the embodied subject) are spatially discrete and locomotive (can change their spatial location), Substance is omnipresent. Substance cannot be locomotive since it has nowhere to go. It already occupies every spatial location and so cannot change its spatial location by moving to a location it did not previously occupy. Consequently, Substance cannot be perceived according to the standards of *MFNS.*[41]

Although the imperceptibility of Substance poses a serious problem for applying the category of substance to Substance, if the category of substance does have application only to Substance, then how could the subject ever cognize relatively enduring empirical objects, or what I have called "substances"? Even though ordinary empirical objects are not sempiternal, they are at least relatively enduring or persistent and so serve substantively in experience.[42] Take Kant's own example, mentioned above, of a piece of wood perishing through incineration while smoke and ash arise. Although Kant uses this example to motivate the idea that there must be something sempiternal that persists through the arising/perishing of these empirical objects, one must also be able to make sense of the alteration of these empirical objects when they are neither arising nor perishing. For example, as the freshly cut wood weathers it changes color. If the category of substance has application only to Substance, however, then how could the subject ever cognize this piece of wood as a relatively enduring empirical object that is suffering alteration of its color? To put it slightly differently, the *scope* of the category of substance seems too great if it applies only to Substance.

Given the scope of the category of substance under the Substance interpretation, how could the subject individuate relatively enduring empirical objects from one another?[43] Particular empirical intuitions surely do not all directly or immediately refer to the same Substance. Again, empirical intuitions are supposed to be singular. Presumably, there is a sense in which the object of one empirical intuition does not refer to the object of another. If the Substance interpretation is correct, however, then every subject-term is ultimately applied to the same Substance and the category of substance has application only to this Substance. How could the category of substance ever have an individuating or reference-dividing use if it has no proper object of application in empirical intuition? As the weathering of wood and examples like it make clear, however, it seems to be a manifest and even necessary feature of our experience that there are many real *substances*.

It does no good at this point to say that one can hold on to substances while neither identifying them with empirical objects on the one hand, nor collapsing them into one big Substance on the other.[44] Assuming that the category of substance applies to substances, but these substances are sempiternal and not relatively enduring, then how could the category of substance ever be applied to these sempiternal substances if relatively enduring empirical objects are all that ever appear in empirical intuition?

Although these are philosophical reasons for adopting a substances interpretation, there are some good textual reasons for adopting this interpretation as well. One textual issue concerns how ordinary empirical objects seem to function as substances in the other two Analogies. Although there are many examples, I will mention only one from the Third Analogy. There Kant intends to establish the principle that 'all substances, insofar as they can be perceived in space as simultaneous, are in thoroughgoing interaction' (*CPR* B256). The example that Kant gives to illustrate this principle is of the earth and the moon standing in causal community with one another. Just as with the burning wood of the First Analogy, it seems that one could experience the annihilation of either the earth or the moon. Notwithstanding their size, they are still merely empirical objects. Given what Kant says about the incineration of wood, however, should this imply that the earth and the moon are not substances? If so, Kant's example in the Third Analogy seems like a non-starter and, furthermore, it is difficult to conceive what a good example would be without recourse to a substances interpretation. If one thinks that the three Analogies stand or fall together, as most commentators do, this is a serious problem that Kant must address.[45]

Although I have just given some textual evidence from the Third Analogy in favor of a substances interpretation, this is not to say that there is no textual evidence for such an interpretation in the First Analogy. As mentioned at the outset of this chapter, it seems that the A-edition statement of the principle of the First Analogy is consistent with ordinary empirical objects serving as substances. It is also important to note Kant's continual use of the locution 'substances' in the First Analogy. For example, in a quote given above (§1.1), Kant says that *substances* are the substrata of all time-determination (*CPR* A188/B231). If what I have said above is correct, it makes little sense to talk about "substances" as opposed to "Substance" if one does not admit that the former refers simply to relatively enduring empirical objects which can appear to subjects in empirical intuition.

My view on substances is in direct conflict with Westphal's understanding of the Analogies. Although parts of the First Analogy are certainly consistent with Westphal's position, it is unlikely that the Analogies can be alternatively described using only the concept of Substance given the nature of human intuition. If relatively enduring empirical objects are all that ever appear in empirical intuition, how can the category of substance ever apply to Substance within the context of the Analogies? Westphal's view, however, requires such an application since without it he cannot make sense of the Analogies using only the concept of Substance. Although Westphal wants to defend common sense causal judgments, such judgments require the union of concepts and intuitions. If empirical intuition offers only relatively enduring empirical objects, causal judgments could only ever be about substances and never about one sempiternal and omnipresent Substance. Even assuming the limitation of causal judgments to substances, however, this does not obviate Kant's need for Substance. In contrast to Westphal, the problem is not that the three Analogies can be alternatively described using only one concept of Substance. Rather, Kant needs both a concept of Substance and a concept of substances in the Analogies, but has recourse only to the one category of substance which seems simply to have been used equivocally in the Analogies.

Returning to the Third Analogy, even if one admits that the earth and moon could not arise or perish *absolutely*, given Kant's arguments in the First Analogy, this should not entail that the earth and moon are mere *states* of Substance and not *substances* in their own right. The former would be close to Westphal's mode-based interpretation of the Third Analogy but would leave little room for the causal community that Kant's account seems to require.[46] If substances are mere states of Substance and states cannot stand in causal community with one another, then there would be no way of distinguishing between the objective succession and simultaneity of these substances given what Kant says in the Third Analogy. This is the mirror-image of the problem that Substance seemed to solve in the Third Analogy. Whereas Substance is necessary for the subject to recognize substances as simultaneous to one another instead of objectively succeeding one another and insures that these substances occupy the same spatial framework, substances must themselves stand in causal community in order to be simultaneous with one another. Westphal's view seems to violate my first principle of interpretation (from §0.1) insofar as it is inconsistent with important texts, but when these texts are taken into account, it violates my third principle of interpretation insofar as it is philosophically incoherent from Kant's perspective.

At this point, it should be clear that when Kant talks about substance he is not always talking about the same thing. There is the relative persistence of individual substances as well as the sempiternal persistence of omnipresent Substance. As I will explain in the next section, this creates a dilemma for Kant's theory of substance that he is unable to resolve within the context of *CPR*. Consequently, this dilemma opens up a gap in the transcendental part of his metaphysics of nature. In subsequent chapters, I will argue that Kant can overcome the above dilemma that faces the Analogies by deducing *a priori* a concept of Substance different from the category (*a priori* concept) of substance that has application only to the substances that appear in empirical intuition. Before moving on to this solution, however, in the final section of this chapter I will examine and criticize three other arguments for something very much like Substance from the Critical period.

1.3 KANT'S DILEMMA AND THE GAP PROBLEM

Kant claims at one point that 'the proposition that substance persists is tautological' (*CPR* A184/B227). This reflects his definition of the schematized category of substance as 'persistence of the real in time' (*CPR* A143/B183). As I have argued, however, Kant does not always mean the same thing when he uses the term "substance."[47] On the one hand, Substance is singular, sempiternal, and omnipresent. On the other hand, there are many individual substances that are only relatively enduring and exist in discrete spatial locations. Although both Substance and substances persist, they persist in very different ways and more than one *a priori* concept of substance is required in order to disambiguate Kant's *different* and mutually *irreducible* uses of "substance."[48]

By way of illustration, consider the following example. Monotheists claim that God is quite powerful while polytheists say the same about their gods. One might even say their respective claims are tautological given the concepts that they are using. Even so, there is a difference in the degree of power one comprehends in the concept "God" (omnipotent) and the concept "gods" (more powerful than humans but less than omnipotent). Likewise, monotheists would not claim that God is somehow reducible to a number of different gods and polytheists (generally) would not claim that their gods are all parts of a single God. The English language has noted the difference between the two concepts through a difference in punctuation and number in the terms used to denote the concepts (they are capitonyms). Just as one concept of god is insufficient to comprehend the difference in power outlined above, so too is the schematized category of substance insufficient to comprehend the difference in persistence in the Analogies.

Likewise, just as the concepts "God" and "gods" are different and mutually irreducible, so too are the concepts "Substance" and "substances."[49] On the one hand, Kant must hold that relatively enduring empirical objects are *substances* since they persist through the alteration of their properties and stand in causal relations with one another. On the other hand, Kant requires one sempiternal and omnipresent *Substance* whose sempiternality insures there is no *absolute* arising or perishing of substances from or to an empty time since either would violate the empirical unity of time. The sempiternality of Substance also safeguards Kant's causal principle from the Second Analogy which requires that no event follows upon an empty time. The omnipresence of Substance in the Third Analogy insures that substances occupy a common spatial framework in which they can causally interact. It is also necessary for the empirical unity of spatial experience and insures that one can distinguish between the objective succession and simultaneity of substances. Consequently, Substance serves as the sempiternal and omnipresent backdrop for the empirical unity of spatiotemporal experience.

Within the context of the Analogies, however, Kant has recourse only to the category of substance and this generates a dilemma for his theory of substance. If the category of substance applies to Substance, then although the category would capture that single backdrop by reference to which the subject experiences the simultaneity and succession of empirical objects in a common spatiotemporal framework, the subject could not individuate relatively enduring empirical objects and perceive their alterations. In contrast, if the category of substance applies to substances, then although the subject could individuate substances and perceive their alterations, the category would not pick out a single backdrop by reference to which these substances could be experienced as either simultaneous with or successive to one another in a common spatiotemporal framework.

Any unequivocal interpretation also fails to make sense of Kant's apparent need for both Substance and substances in the other two Analogies. Insofar as the application of the category of substance must be limited to substances, Kant requires a new *a priori* concept of Substance different from the category of substance. In my view, this constitutes a gap in Kant's Critical philosophy that he is unable to deal with given the tools available to him in *CPR*. This is a gap in the transcendental part of the metaphysics of nature since it lies at the heart of his project in the Transcendental Analytic, *viz.* the Analogies of Experience.

In *Farrago* 1 of *OP*, which was written in December 1798 not long after his fall 1798 letters to Garve and Kiesewetter, Kant says that the transition project 'is designed to fill what is still a gap in the pure doctrine of nature and generally in the system from *a priori* principles, and so toward accomplishing completely my metaphysical work' (*OP* 21:626.8–11), and Kant suggests, on this same leaf, that his *a priori* concept of the ether will be instrumental in filling this gap.[50] This quote is consistent with my interpretation insofar as the "transcendental part" of the metaphysics of nature consists of a 'system from *a priori* principles' within Kant's 'pure doctrine of nature.' The *a priori* concept of the ether would be an important addition to the system from *a priori* principles that could complete Kant's pure doctrine of nature.[51]

If one accepts my interpretation, it is the post-Critical development of Kant's *a priori* concept of the dynamic ether that provides him with the conceptual resources necessary to bridge the gap I have identified in the Critical philosophy since it performs all the functions that the concept of Substance must perform in the Analogies while at the same time being importantly different from Kant's schematized category of substance in *CPR*.[52]

One should note that, given the nature of empirical intuition, the category of substance is best understood as having application only to substances. Although Westphal believes that the three Analogies can be alternately described using only the concept of Substance, it is doubtful that the category of substance can even apply to Substance if only substances appear in empirical intuition.

This does not resolve the dilemma so much as it illustrates the high costs of adopting only the substances horn. If the category of substance has application only to substances, this calls into question Kant's arguments in the Analogies which certainly seem aimed at establishing something like Substance. If Kant's arguments in the Analogies fail to prove Substance, however, he requires a new argument for Substance. I believe that Kant provides such an argument in the Ether Deduction. By proving the *existence* of Substance while also deriving its concept *a priori,* Kant's post-Critical philosophy has the resources necessary and sufficient to bridge the gap in his Critical philosophy, i.e., to complete the system of *a priori* principles left incomplete in the Analogies. In fact, Kant seems to recognize the need for such an *a priori* existence claim in the Refutation of Idealism, though he does not provide a proof for it there:

> Not only can we perceive all time-determination only through the change in outer relations (motion) relative to that which persists in space (e.g., the motion of the sun with regard to the objects on the earth); we do not even have anything persistent on which we could base the concept of a substance, as intuition, except merely *matter,* and even this persistence is not drawn from outer experience, but rather presupposed *a priori* as the necessary condition for all time-determination, thus also as the determination of inner sense in regard to our own existence through the existence of outer things.
>
> (*CPR* B277–278)

Although perceiving the motion of substances in outer sense is necessary for time-determination in inner sense, these substances do not persist in the sense that Substance (matter) is persistent. We must presuppose the latter *a priori* as underlying substances and, by extension, time-determination in inner sense. Even though this anticipates what Kant has to say post-Critically, the Ether Deduction will provide a *proof* for this Critical-era presupposition.

Up to this point, my focus has been on Kant's attempt to prove the existence of Substance using the category of substance in the Analogies. Perhaps there is another argument, however, for Substance outside of the Analogies that will allow Kant to overcome the above dilemma. Before moving on to what I consider to be Kant's solution to the dilemma in *OP,* I would first like to examine three other arguments for something very much like Substance from the Critical period. If Kant is able to establish the existence of Substance using only the conceptual resources available to him in the Critical period, then he may not have to make appeal to a post-Critical solution. I will argue, however, that each of these arguments is either unsound or unable to establish the kind of conclusion that Kant needs to avoid the dilemma that faces his theory of substance in the Analogies.

First Alternative Argument for Substance: Transcendental Ideal of *CPR*

Outside the Analogies, there is another important reference to a notion equivalent to that of Substance in *CPR*. In the Transcendental Ideal, Kant claims that 'nothing is an object for us, when it does not presuppose the totality of all empirical reality as a condition of its possibility' (*CPR* A582/B610). Similarly, Kant says that 'the matter for the possibility of all objects of sense must be presupposed as given in one totality' (*CPR* A582/B610). I believe that this "totality of empirical reality" or "matter" is equivalent to his conception of Substance. Kant explains his view using what he calls the 'principle of thoroughgoing determination' (*CPR* A571/B579).[53] This principle states that for any one place predicate and for any given empirical object either the predicate or its opposite must apply to the object. We must think that *every* empirical object stands under this principle as a condition of the object's *possibility.* As Kant says:

> Every thing, however, as to its possibility, further stands under the principle of thoroughgoing determination; according to which among all possible predicates of things, insofar as they are compared with their opposites, one must apply to it.
>
> (*CPR* A571–572/B599–600)

The principle is not merely a restatement of the law of the excluded middle (which is an analytic law of pure general logic), however, insofar as the principle presupposes the 'sum total of all predicates of things in general'

(*CPR* A572/B600) so that the predicates of the object might be determined relative to such a sum total. This sum total is not given in the principle itself and so renders the principle synthetic in its application. As Kant says:

> The principle of thoroughgoing determination thus deals with the content and not merely the logical form. It is the principle of the synthesis of all predicates which are to make up the complete concept of a thing, and not merely of the analytic representation, through one of two opposed predicates; and it contains a transcendental presupposition, namely that of the matter of all possibility, which is supposed to contain *a priori* the data of the particular possibility of everything.
>
> (*CPR* A572–573/B600–601)

Although Kant believes that the principle of thoroughgoing determination can legitimately apply only to empirical objects, they will nonetheless presuppose a unified whole, or totality, which contains the possibility of their thoroughgoing determination. This is a totality relative to which the predicates of an empirical object might be determined. In fact, Kant holds that 'empirical objects, their difference from one another and their thoroughgoing determination, can rest only on the limitation of this sum total' (*CPR* A582/B610). The totality itself, however, is not *subject* to the principle of thoroughgoing determination even if it is *presupposed* for the application of the principle to empirical objects. Empirical objects are specific determinations (or limitations) of a totality which is itself radically undetermined. As we will see in the next chapter (§2.2), Kant says similar things about the ether or Substance in *OP*. Although the ether or Substance is characterized *negatively* relative to the properties that substances possess, its actuality is nonetheless necessary for the *positive* determination of substances.

Just as the principle of thoroughgoing determination has a *legitimate* application, it also has an *illegitimate* application, one that ties the principle to Kant's general aim in the Transcendental Ideal, *viz.* exposing the errors of rational theology. Although it is legitimate to presuppose a 'distributive unity of the use of the understanding' (i.e., a unity produced by the synthetic activity of understanding) as a condition for the thoroughgoing determination of any *empirical object,* we cannot transform this distributive unity into a 'collective unity of a whole of experience' and subsequently 'hypostatize' or assert the existence of an 'individual thing containing in itself all empirical reality' as a condition for the thoroughgoing determination of *things in general* (*CPR* A582–583/B610–611).[54] For rational theology, this individual thing is God, the *ens realissimum* or most real being.[55]

Although the illegitimate use of the principle clearly commits one to an objectionable form (at least from Kant's perspective) of transcendental realism, is the legitimate use of the principle consistent with Kant's transcendental idealism/empirical realism? Assuming that the 'totality of empirical reality' can only be legitimately understood as a 'distributive unity of the use of the understanding,' the totality presupposed by the principle of thoroughgoing determination would itself be a product of the understanding. This is at best circular, and at worst would commit Kant to some form of empirical idealism. Given that perceptions require that the subject be affected by *actual* objects of experience in sensibility, the aggregation of perceptions cannot itself *produce* the totality of empirical reality, since each perception *assumes* such a totality for the mere *possibility* of its object.[56] Put in terms of the principle of thoroughgoing determination, the totality of empirical reality is *assumed* for the thoroughgoing determination of an actual object and so cannot be the *product* of those determinations aggregated together. On the other hand, if the understanding produces the totality of empirical reality as an antecedent condition for the determination of empirical objects, then the understanding would provide both the 'material for the possibility of all objects of sense' (*CPR* A582/B610) as well as the form of these objects, which would seem to commit Kant to empirical idealism since empirical objects would simply be the product of mental activity.[57]

One might object that I am holding Kant's *presupposition* of the totality of empirical reality to an unreasonable standard. The objector might hold that Kant's presupposition is only *regulative* whereas I am insisting that it be *constitutive.* Although I am claiming that the totality of empirical reality must *exist,* perhaps Kant is only claiming that we must approach the determination of empirical objects *as if* the totality of empirical reality exists. In this respect, Kant's presupposition of the totality of empirical reality for theoretical purposes is no different than his presupposition of God, the soul, or human freedom for practical purposes. I do not think a regulative presupposition is strong enough for Kant's theoretical purposes. Although it is easy enough for me to act *as if* God exists for practical purposes, it would be *impossible* for the computer in front of me to be an object of experience if the totality of empirical reality did not exist. Granting that the computer in front of me is an actual object of experience, it presupposes constitutively the totality of empirical reality. Whereas problematic concepts like God, the soul, or freedom are regulatively necessary to insure the possibility of their respective objects, the totality of empirical reality must be constitutively necessary to insure the actuality of the objects of experience. The objects of experience are not merely possible (like God, the soul, or freedom), but rather actual. Consequently, the totality of empirical reality must serve a constitutive function even within the context of the Transcendental Ideal.[58]

The totality of empirical reality can neither be a product of the understanding (circular or commits Kant to empirical idealism) nor can it be a thing-in-itself (e.g., God, which commits Kant to transcendental realism). In the Transcendental Ideal, Kant fails to provide an argument for the totality of empirical reality that is consistent with his transcendental idealism/empirical realism and has the constitutive force that he requires. Even if Kant fails to provide this argument in *CPR*, however, he does offer such an argument in the Ether Deduction of *OP*.[59] As I will argue, the conclusion of the Ether Deduction is not only a constitutive thesis but also wholly consistent with Kant's transcendental idealism/empirical realism.

Second Alternative Argument for Substance: Dynamics of *MFNS*

One might claim that in Proposition 4 of the Dynamics section of *MFNS* Kant offers a different argument for something very much like Substance. At first blush, the proof for this proposition seems to offer an argument for omnipresent matter based upon Kant's argument for the infinite divisibility of matter.[60] Kant argues that space is infinitely divisible, though he believes that this argument cannot establish that matter is infinitely divisible unless is has been shown that 'there is material substance in every part of space' (*MFNS* 4:504). Although one might now expect Kant to offer an argument that there is material substance in every part of space, what he actually argues is that every *filled* space is infinitely divisible. This does not entail, however, that every *space* is filled. Kant ties the mathematical argument for the infinite divisibility of space to his metaphysical account of how matter is constituted dynamically. He even uses the monadic view (which typically rejects the infinite divisibility of matter but is committed to the existence of empty space) to illustrate his point. According to Kant, even if matter were constituted out of physically indivisible parts (monads), they would nevertheless need to occupy space dynamically. For Kant, space is filled by repulsive force and every point within a sphere of repulsive force (e.g., point 'c') will likewise exert its own repulsive force against any point at the surface of the sphere (e.g., point 'a') as well as to the point at the center of the sphere of repulsion (e.g., 'A'). The same argument can be run again for some point between 'c' and 'A' (or between 'a' and 'c') *ad infinitum*. Since '*repelling* is a [kind of] moving, c is something movable in space, and thus matter' (*MFNS* 4:504). Consequently, for any given matter that fills space (through repulsive force), it will be infinitely divisible into other matters that fill space (through repulsive force). The repulsive force that fills a space is *continuous* rather than *discrete* and so infinitely divisible.[61] This is consistent with Kant's definition of a "quantum continuum" (no part is simple) in the Anticipations of Perception in *CPR*.[62] There he claims that *all* appearances in general are continuous magnitudes.[63] In *MFNS*, force, movability, and divisibility are all continuums.[64]

A few pages later, in Proposition 7 of the Dynamics section of *MFNS*, Kant makes clear that he does not think that every part of space is filled with matter. He defines the original attraction essential to all matter (in conjunction with original repulsion) as the 'immediate action of matter on other matter through empty space' (*MFNS* 4:512). Space cannot be full of matter, according to Kant, or we would not be able to talk about *one* matter acting on *another* matter. If one matter could not be outside of another (i.e., separated by empty space), one matter could not act immediately on another except through impenetrability (original repulsion).[65] This would undermine Kant's dynamical theory which relies on *two* fundamental forces. Attractive force would be 'either completely impossible or always dependent on the action of repulsive forces' (*MFNS* 4:513). In this section, Kant even discusses his example from the Third Analogy, holding that the space between the earth and the moon 'may well be viewed as completely empty. For even though matter may lie between the two bodies, it still contributes nothing to this attraction' (*MFNS* 4:513).

This might seem to contradict what Kant says in the Third Analogy where he insists that "matter everywhere" is necessary for causal community. Something can be a necessary condition for a causal interaction, however, without contributing causally to that interaction. In the Third Analogy, "matter everywhere" insures that the earth and the moon share the same spatial framework and so are not causally isolated from one another. Although "matter everywhere" provides the appropriate backdrop for causal community, the causal community itself depends upon the causal powers of the substances that stand in causal community (i.e., the earth and the moon).[66] This strongly suggests that relatively persistent and spatially discrete empirical objects (like the earth and the moon) should not be understood simply as *properties* of Substance, but must rather be understood as ontologically robust *substances* in their own right. Kant's discussion in *MFNS* does raise the question, however, of how we should understand "empty space." What I believe Kant is saying, to use my terminology, is that space is not full of *substances.* Space can be full of *Substance,* though only as long as the concept "Substance" is not understood as having the same extension as the concept of "substances." For example, in *MFNS,* Kant suggests the possibility of an ethereal medium that occupies space dynamically (Substance) even if it leaves space mechanically empty, i.e., allows for the free locomotion of bodies (substances) within it.[67] One should not think, however, that Kant takes himself to have established the existence of anything omnipresent in *MFNS.* In fact, in the General Remark to the Dynamics, Kant views omnipresent ether as a *hypothesis* that provides a thinkable alternative to the atomist's hypothesis of empty space.[68] If Kant thought that he had already established the existence of something omnipresent in the main part of the Dynamics, it is unclear why he would view it as a hypothesis in the General Remark to the Dynamics. Kant's discussion of the ether in *MFNS* also suggests that he likely viewed his arguments in the Analogies as only establishing the existence of substances but not Substance. Although I have argued that this is the most that the Analogies can establish, it in no way obviates Kant's need for Substance given the dilemma that his theory of substance faces as well as the arguments of §1.1. Even though it cannot resolve the dilemma, I will discuss Kant's conception of the ether in *MFNS* at greater length in the next chapter (§2.1).

Third Alternative Argument for Substance: Dynamics of *MFNS* and Early *OP*

Notwithstanding the problems with the two arguments discussed above, there is a third argument that may offer Kant a way out of the dilemma that his theory of substance faces but that requires only one concept of substance. As we will see, this argument has the additional advantage of *bridging* Kant's discussion in *MFNS* and his discussion in *OP*. Consequently, it might come closest to offering Kant's considered opinion on this issue. Using the empirical concept of matter (the moveable in space), Kant argues in *MFNS* both that (1) the matter in space is itself substance, and (2) the parts of matter insofar as they are subjects and not mere predicates of one another are substances as well. As Kant says:

> Thus matter, as the movable in space, is the substance therein. But all parts of matter must likewise be called substances, and thus themselves matter in turn, insofar as one can say of them that they are themselves subjects, and not merely predicates of other matters.
>
> (*MFNS* 4:503)[69]

If one identifies the "matter everywhere" in the Third Analogy with the "matter" he mentions here in *MFNS,* both could be seen as referring to Substance. Likewise, substances could be viewed simply as parts of Substance. Consequently, Kant would require only one concept of substance which itself applies to Substance, but in conjunction with a "part of" operator can be seen as applying to substances as well. Insofar as this matter is omnipresent and sempiternal, it seems as if this interpretation could do the work of Substance in the Analogies. Likewise, the parts of matter could do the work demanded of substances in the Analogies.

This solution seems to show up in *OP* as well. In a section of *OP* dating from 1798, Kant reiterates his view from *MFNS:* 'Matter is the movable in space (and in time). This matter, so far as it is limited (through inner attraction), is a body, that is, a whole as *substance* so far as it is movable and moving other matter' (*OP* 21:347.5–8).[70] This view requires only one concept of substance and that is the concept of a sempiternal and omnipresent Substance. Relatively enduring empirical objects, or what I have called "substances," are simply *parts* of Substance. They are temporary configurations of something which is itself sempiternal and can be distinguished from one another on the basis of their relative motions. Under this view, the category of substance can apply to Substance as well as to substances in the same way as the concept of a mass-term like "water" might be applied to the whole of water or any of its macrophysical parts. There is no property that the latter might possesses that the former could not and vice versa. At every macrophysical level of division, you still have *water.* If the above proposal is correct, much the same would be true of the relationship between

Substance and substances. At every level of division, you still have sempiternal matter. This view is *reductive* insofar as there is no essential difference between the properties of Substance and substances. Even if Substance as a whole is never present to empirical intuition, one can still claim cognition of Substance through its parts that do appear in empirical intuition, i.e., substances. Consequently, only one concept of substance is required in order to capture both Substance and substances.[71]

It seems to me, however, that the indifferent denotation of mass-terms only tolerates difference in *quantity* (amount), but not in *quality.* With regard to Substance vs. substances, however, the qualities also seem different and mutually exclusive, e.g., sempiternality vs. relative persistence. It is also difficult to see how the properties of matter in *MFNS* (e.g., motion, impenetrability, inertia) can be attributed to Substance. One might respond by saying that substances are themselves just *properties* of Substance. Although Substance is *essentially* sempiternal and omnipresent, substances are simply *accidents* of Substance at certain places and times. Put differently, the spatiotemporally discrete configurations of Substance (i.e., substances) are the only things that are relatively persistent though the matter configured is sempiternal. Under this view, however, there is no principled difference between the alteration and the arising/perishing of substances since both would have to be explained simply in terms of reconfiguration of Substance.[72] I believe there is an important distinction to be drawn, however, between the alteration of substances (e.g., leaf changing color) and the alteration of Substance (e.g., burning wood changing into smoke and ash). The alteration of states of Substance involves the replacement change of substances, whereas the alterations of states of substances involve non-substantial replacement changes. The alteration of Substance involves a change of states that could not occur at the level of substances and vice versa.

There is also a metaphysical concern if substances are just accidental properties of Substance. If substances are properties of Substance, how can they support their own properties? This seems like a category mistake. As mentioned above, given Kant's views on causal community, can *properties* stand in relationships of causal community to one another? If we take Kant at his word, *substances* not the *properties* of substances stand in causal community (e.g., the earth and the moon in the Third Analogy). Properties do not themselves possess causal efficacy, but rather the substances that support those properties.[73] Kant's comments in Proposition 7 of the Dynamics of *MFNS* (discussed above) seem to support this interpretation of substances as well.

Most importantly, although Substance can arguably fall under the concept of matter in *CPR* (see §1.1), it cannot fall under the concept of matter in *MFNS.* In the above quote from *MFNS,* Kant holds that 'matter, as the movable in space, is the substance therein.'[74] He reiterates this in the above quote from *OP,* saying that 'matter is the movable in space (and in time).' In *MFNS,* Kant says the 'motion of a thing is the *change of its outer relations* to a given space' (*MFNS* 4:482), i.e., motion is *locomotion.*[75] Speed and direction are the two basic properties of motion.[76] Something becomes an object of experience 'when a certain *object* (here a material thing) is thought as *determined* with respect to the *predicate* of motion.' As omnipresent, however, speed and direction cannot be predicated of Substance. The latter is not locomotive and so cannot be an object of experience.[77] Substance does not stand in the right kinds of external relations to affect the senses as substances can.[78] Within the context of *MFNS,* it is unclear whether Substance even falls under Kant's definition of "nature," which is 'the sum total of all things, insofar as they can be *objects of our senses*' (*MFNS* 4:467). Given the nature of Substance, it is better to understand Kant's discussion of "matter" and "body" in *MFNS* as referring simply to substances.[79] As we will see in the next chapter, after 1798, Kant comes to rethink his view on matter, identifying it with something much closer to Substance.

Assuming that substances do possess some properties or states that are irreducible to the properties or states of Substance, I will argue in the next chapter (§2.2) that substances *emerge* from but are not *reducible* to Substance.[80] Although this is a contemporary philosophical term not used by Kant, I think that it does the best job of clarifying what is otherwise a murky discussion in *OP.*[81] As I will argue, the emergence relation is tracked through the application of two separate and irreducible concepts. One concept operates at the level of substances, whereas the other concept operates at the level of Substance. This view is very different, however, from the proposal sketched above. If one takes substances simply as *parts* of Substance, then substances would be reducible to Substance in the same way as any macrophysical part of water is still water. Although already problematic within the context of the Analogies, the view becomes unsustainable once one views Substance as a dynamic force plenum, which Kant does later in *OP.*[82] As mentioned above, this dynamic force plenum, or what Kant calls "ether," does not have any of the determinate physical properties that substances possess but is rather that which makes these determinate physical properties themselves possible. In addition, this plenum is a *continuum* of forces while substances are necessarily *discrete.* Viewing substances simply as parts of Substance does not seem consistent with the post-Critical developments in Kant's theory of matter.

* * *

Even if the various solutions to the dilemma offered above must be abandoned, I believe that a solution to Kant's problem can be found by examining the post-Critical development of Kant's conception of Substance in *OP.* The following chapters will discuss these developments. The next chapter will track the development of Kant's ether theory from the Critical period to the post-Critical period. Chapter Three will describe Kant's attempt to deduce *a priori* the existence of Substance, or what he comes to call the "ether" in the Ether Deduction of *OP.* This deduction will provide the argument for Substance that is simply presupposed in the Transcendental Ideal section of *CPR* and which the First Analogy is ill-suited to provide given the likely application of the category of substance simply to substances. Given that Kant's proof strategy is *a priori,* he will claim that the concept of the ether is itself *a priori.* I will argue below that this *a priori* concept of the ether is the concept of Substance. If Kant's deduction is successful, he will not only have proven the existence of Substance, but will have also added an *a priori* concept of Substance different from the category of substance. I will argue in Chapter Four that having these *two* concepts in hand allows Kant to avoid the dilemma that faces him in the Analogies and so bridge the gap in his Critical philosophy.

NOTES

1. Veit-Justus Rollman and Andree Hahmann also recognize that Kant equivocates on the meaning of "substance" between the A- and B-edition versions of the Principle of the First Analogy. They claim, however, that the B-edition principle is meant to replace the A-edition principle. They go on to argue for a Substance interpretation of the First Analogy while recognizing that this leaves Kant with the problem of explaining substances. In what follows, I will try to provide a solution to this problem. See Veit-Justus Rollman and Andree Hahmann, "Weltstoff und absolute Beharrlichkeit: Die Erste Analogie der Erfahrung und der Entwurf Übergang 1–14 des Opus postumum," *Kant-Studien* 102, no. 2 (2011): 170–171.
2. For the former see *CPR* A80/B106. For the latter see *CPR* A70/B95.
3. See *CPR* A79/B105.
4. Whereas empirical intuitions are related to their objects through sensation and are *a posteriori,* pure intuitions are *a priori* and nothing is to be encountered in them which belongs to sensation. See *CPR* A20/B34.
5. See *CPR* A144/B183. See also *CPR* A177/B219 for Kant's description of the three *modi* of time in the Analogies.
6. Unless otherwise noted, when I refer to the category of substance, I am referring to the category in its schematized form. It might be tempting, however, to resolve the above dilemma by using Kant's distinction between the schematized and unschematized forms of the categories. One could claim that whereas the schematized category of substance has application to substances, the unschematized category of substance has application to Substance. This would account for both substances and Substance using only the single category of substance. One must remember, however, that the First Analogy, under both the Substance and substances interpretations, has to do with *persistence* (schematized category) and not merely the substance/attribute relation (unschematized category). For Kant's description of the unschematized use of the categories see *CPR* A146–A147/B186. In addition, Kant makes clear in *CPR* that something falling under the unschematized category of substance is insufficient to assert its real existence. See *CPR* B407.
7. See *CPR* A161/B200.
8. Henry Allison, *Kant's Transcendental Idealism: An Interpretation and Defense,* 2nd ed. (New Haven: Yale University Press, 2004), 240.
9. See *CPR* A185/B228.
10. Allison, *Kant's Transcendental Idealism,* 237–239. Jonathan Bennett has argued that this kind of reasoning involves a non-sequitur, and more precisely,

a quantifier-shift fallacy. Even if throughout *any* replacement change there is something or another which remains in existence, this does not entail that there is some one thing which remains in existence throughout *every* replacement change. Again, there is no reason to think that there is one Substance that persists through every replacement change just because there is some substance or another that persists through any particular replacement change, e.g., wood perishing into smoke and ash. See Jonathan Bennett, *Kant's Analytic* (Cambridge: Cambridge University Press, 1966), 199–200. Allison attempts to overcome the non-sequitur Bennett mentions by switching the focus of Kant's argument from a logical or ontological claim to an epistemic one. Instead of taking Kant's claim to involve the idea that something simply arising/perishing is itself *self-contradictory*, he takes the idea to be that something arising/perishing without some underlying object that persists could not be an object of possible *experience*. Allison also wants to defend an epistemic reading of sempiternality as a condition of unitary experience in a single time. See Allison, *Kant's Transcendental Idealism,* 242–244. Paul Guyer agrees with Allison that, for Kant, the only true substance would be a sempiternal substance since no possible experience might count as verification of the annihilation of a genuine substance. See Paul Guyer, *Kant and the Claims of Knowledge* (Cambridge: Cambridge University Press, 1987), 216 and 231. As I will argue in this chapter, Kant has good epistemological arguments (having to do with the unity of our spatiotemporal experience) as well as good metaphysical arguments (having to do with the nature of causation) for affirming the backdrop thesis. In neither case, however, will Kant's arguments trade on the quantifier shift that Bennett identifies.

11. The idea that the only genuine kind of substance is sempiternal seems to be the received view in the literature. Bennett distinguishes between two kinds of substance that show up in the First Analogy: substance$_1$ and substance$_2$. Whereas the former defines "substance" as a bearer of properties, the latter defines "substance" as something sempiternal. As mentioned above, however, Bennett believes that Kant's argumentative move from substance$_1$ to substance$_2$ involves a non-sequitur. Although Bennett challenges Kant's *argument,* he is sympathetic with its *conclusion.* Bennett believes that substance$_2$ is the only kind of substance for Kant, though it is practical to retain the conception of substance$_1$ for linguistic economy. One must understand, however, that substance$_1$ exists only adjectively upon substance$_2$. See Bennett, *Kant's Analytic,* 182–184, 197–198. P.F. Strawson seems to agree that, for Kant, relatively enduring substances are so only by "courtesy" since they are adjectively substantial. See P.F. Strawson, "Kant on Substance," in *Entity and Identity and Other Essays* (Oxford: Oxford University Press, 1997), 269–270. For his part, Arthur Melnick agrees with the non-sequitur that Bennett mentions while accepting Kant's conclusion. See Arthur Melnick, *Kant's Analogies of Experience* (Chicago: Chicago University Press, 1973), 67–71. Although I have already mentioned Allison and Guyer in connection with Bennett, James Van Cleve combines Bennett's conceptions of substance1 and substance$_2$ when he defines "substance" as the concept of something that exists only as subject and at all times. Like Bennett and Strawson, Van Cleve claims that ordinary empirical objects exist only adjectivally as ways in which the sempiternal exists. According to Van Cleve, however, these empirical objects (phenomena) are virtual objects or logical constructions by perceivers where the latter are noumenal beings and the only things that could qualify as genuine substances. See James Van Cleve, *Problems from Kant* (Oxford: Oxford University Press, 1999), 106 and 121. Andrew Ward agrees with the above commentators that the only genuine kind of substance is sempiternal. Whereas some of the

above commentators oscillate between the omnipresent (Substance) and individuating use of "substance," Ward argues for a plurality of sempiternal and individuated substances. See Ward, "Kant's First Analogy of Experience," *Kant-Studien* 92, (2001): 387–406. Westphal goes the other way, arguing that the three Analogies can be alternatively described using only one sempiternal Substance if substances are just modes of Substance. See Westphal, *Kant's Transcendental Proof of Realism,* 147–166. Like Westphal, Rollmann and Hahmann suggest that substances are just accidents of Substance. See Rollmann and Hahmann, "Weltstoff und absolute Beharrlichkeit," 188.

12. Although Strawson agrees with the idea that there must be a common temporal framework, he does not believe that such a framework requires anything sempiternal. For example, a common time could be established simply on the basis of temporally *overlapping* relatively enduring substances. If substance A overlaps temporally with substance B and substance B overlaps temporally with substance C, substance A might perish by the time substance C arises without violating the unity of time since substance B overlaps temporally with both substances A and C. Such a framework of overlapping substances would establish a common time, though no particular substance within the framework would need to be sempiternal. See Strawson, *The Bounds of Sense: An Essay on Kant's Critique of Pure Reason* (London: Methuen & Co., 1966), 128–130 and "Kant on Substance," 274. In assuming that these substances overlap in time, Strawson is assuming that they occupy the same time. As I have argued, given the fact that each absolutely arising substance comes from a time that does not contain any of the other substances, however, there is no reason to assume that they would occupy the same time. Although Kant does not have a problem with *substances* overlapping in time (as long as they do not arise or perish absolutely), he does have a problem with *times* overlapping.
13. See also *CPR* A186/B229 for more on the relationship between the sempiternality of substance and the unity of time.
14. Allison offers a somewhat different defense of the conservation of substance thesis based on the idea that the persistence of substance could only be understood in terms of its *quantity.* See Allison, *Kant's Transcendental Idealism,* 244–246.
15. See *CPR* B233.
16. As mentioned above, this is Ward's view. See Ward, "Kant's First Analogy of Experience."
17. Kant holds that there is no experience of empty space or empty time elsewhere during the Critical period. See *CPR* A172/B214, A487/B515, A521/B549 and *MFNS* 4:559.
18. Here I am following Eric Watkins. See Watkins, "Kant on Rational Cosmology," in *Kant and the Sciences,* 76–77. For the relevant part of Kant's lectures, see *AK* 29:922–923.
19. Friedman believes that Kant's mention of 'matter everywhere' in the Third Analogy refers to the light-ether necessary for us to be in perceptual contact with heavenly bodies. See Friedman, *Kant's Construction of Nature,* 359. Although I will argue that there is certainly a connection between "matter everywhere" in the Third Analogy and Kant's post-Critical conception of the ether (which is not equivalent to Euler's light-ether), the omnipresence of matter is necessary not only for our perception of the heavenly bodies, but also for the causal community between them.
20. Both Edwards and Westphal spend some time discussing Kant's arguments against empty space in the Third Analogy and how these arguments support an omnipresent and dynamical view of substance. Even so, they fail to detect any tension within Kant's theory of substance. See Edwards, *Substance, Force,*

and the Possibility of Knowledge, 26–43 and Westphal, *Kant's Transcendental Proof of Realism,* 80–82. Robert Hanna spends some time examining the First Analogy, specifically where he recognizes two different uses of "substance." Although he overlooks the connection between what I call "Substance" and Kant's affirmation of "matter everywhere" in the Third Analogy, he does identify what he calls the "One Big Substance" with matter as a whole and views matter via a dynamical theory. Even so, his solution to the dilemma that faces the Analogies is different from my own. I will return to his view in the final section of this chapter. See Robert Hanna, *Kant, Science, and Human Nature* (Oxford: Oxford University Press, 2006), 390–408.

21. See *CPR* B291. As mentioned in the Introduction (§0.4), Förster's interpretation of the gap hinges on the idea that Kant needs a separate spatial schematism of the categories in order to insure their objective reality. If I am right, however, Kant sees no problem with applying the temporally schematized categories to space in *CPR* to insure their objective reality. See Förster, *Kant's Final Synthesis,* 59.
22. See *CPR* A162/B202–203.
23. Emphasis mine. For the extension of the Anticipations to space, see *CPR* A174/B216.
24. See *CPR* A172/B214.
25. See *CPR* A184/B227. Allison does much the same with his backdrop thesis. See Allison, *Kant's Transcendental Idealism,* 240 and 242.
26. Beiser argues that Kant's distinction between the transcendental (*a priori*) and the empirical (*a posteriori*) is one that Kant remained sensitive to throughout both the Critical and post-Critical periods. See Beiser, *German Idealism,* 199.
27. Even if one admits that "matter everywhere" is merely an *empirical* condition in the Third Analogy, it is clear that Kant's argument for the ether in *OP* intends to establish a *transcendental* material condition of experience. I will examine this argument at greater length in Chapter Three. Beiser also endorses the view that Kant's argument for the ether in *OP* is transcendental and not empirical. See Beiser, *German Idealism,* 188.
28. I will also argue that Substance is a condition for the possibility of the objects of experience (i.e., substances), without itself being an object of experience.
29. For similar claims, see *CPR* B218 and A188–189/B232.
30. It is also important to be careful when discussing "metaphysics" within the context of Kant's theoretical philosophy. As I will argue below, my interpretation of Kant (both Critically and post-Critically) does not commit him to any form of transcendental realism, but rather is thoroughly consistent with his dual theory of transcendental idealism and empirical realism (see §3.4).
31. Westphal, *Kant's Transcendental Proof of Realism,* 172.
32. Although Westphal believes that the Analogies require the metaphysical causal thesis, Kant only argues for this thesis explicitly in Proposition Three of the Mechanics chapter of *MFNS* (his Second Law of Mechanics). Westphal holds that this latter argument fails. Specifically, he argues that all changes in matter could be due to a Spinozistic or living matter (hylozoism) capable of producing its own alterations. According to this view, substances are simply modes of this Substance. Although in his 1790 work, *On a Discovery Whereby any New Critique of Pure Reason is to be Made Superfluous by an Older One,* Kant resists characterizing substance in terms of forces for fear of Spinozism (*AK* 8:224), in *OP* Kant sometimes favorably compares his view to Spinoza's view (for example, see *OP* 22:304). For a detailed discussion of these sections of *OP,* see Edwards, "Spinozism, Freedom, and Transcendental Dynamics" and Lord, *Kant and Spinozism,* 175–183. One worry with Westphal's approach is that he might illegitimately import general metaphysics, which is concerned

with all substances in general, into the special metaphysics of *MFNS,* which is concerned solely with non-living material substances. See *MFNS* 4:544 and Friedman, *Kant's Construction of Nature,* 336. In addition, as I will argue below, Kant still has a need for substances as a distinct class of entities that emerge from but are not reducible to Substance. If I am right, this is enough to show that he cannot be committed to full-fledged Spinozism.

33. See Westphal, *Kant's Transcendental Proof of Realism,* 147–166.
34. See *CPR* A320/B377.
35. Emphasis is mine.
36. I will argue in §4.1 that Kant resolves this issue post-Critically in *OP.* There Kant draws a distinction between direct appearances of the ether/Substance and indirect appearances of substances. Whereas the subject has only subjective perceptions of the ether/ Substance (direct appearance), the subject has objective perceptions (indirect appearance) of substances. For Kant's distinction between subjective and objective perceptions in *CPR,* see A320/B376.
37. In this respect, one might compare the role of Substance in the Analogies to the role that absolute space plays in *MFNS.* Although absolute space must be *thought* as the background for the perception of relative motion, absolute space is not perceived and is only a regulative idea (*MFNS* 4:481–482 and 4:560). Even if we must *think* of Substance as grounding our experience of substances, this is not sufficient to make the *existential* claim that Substance (or a backdrop) exists. In the Second Postulate of Empirical Thought, Kant insists that all cognition of existence requires either (1) the perception of what is claimed to exist or (2) the perception of something else, connected in accordance with the Analogies of Experience, with what is claimed to exist (*CPR* A225/B272–273). Although I will argue that we can affirm the existence of Substance using the second strategy, this is not an argument that is based simply upon the necessity of *thinking* Substance. To claim that Substance exists simply because it must be thought of as existing, would be to make an error quite similar to the one that the rational psychologist makes in the Paralogisms with regard to the soul, *viz.* claiming that a substance exists as a real subject of inherence since it must be thought of as existing as a logical subject of predication (*CPR* B409–411). There are, however, parallels between the arguments that Kant criticizes in the Paralogisms and his own arguments in the Analogies. Just as there is no objective perception of the "I think" in the Paralogisms, so too is there no objective perception of Substance in the Analogies. Both arguments also rely on fallacies of equivocation. As noted above, the Paralogisms equivocate between *logical* and *real* subjects or properties. The arguments in the Analogies equivocate between *Substance* and *substances.*
38. For Van Cleve's objections to the First Analogy, see Van Cleve, *Problems from Kant,* 107–108. I will argue below that even if the backdrop (what I call "Substance") cannot be perceived objectively, that nevertheless there must be a *constitutive* use of the concept of Substance. When it comes to the category of substance, however, I will argue that Van Cleve's worry can be avoided since the category of substance applies to substances which can be objectively perceived. In addition to the above criticism, Van Cleve also deploys Bennett's objection to the backdrop thesis discussed above. The arguments I have offered in this chapter, however, can be seen as a way of meeting Bennett's objection. In conjunction with Bennett's objection, Van Cleve also argues that the unity of time explanation is a non-starter since if such a position were correct then something like the ether would be required for the unity of space. As I will argue in Chapter Three, when presenting the Ether Deduction, however, this is *exactly* what Kant means! Finally, Van Cleve argues that, even assuming the existence of the backdrop, substances could still arise and perish absolutely

if they are not alterations of Substance. For Kant, however, the most important point is that a substance cannot come into existence from an *empty* time (since this would violate the unity of time) and a sempiternal backdrop seems sufficient to preclude this possibility. In addition, Kant's dynamic theory of matter as developed in *OP* requires that substances emerge from Substance. I will discuss this view at greater length in the next chapter.

39. See Friedman, *Kant's Construction of Nature,* 40–44.
40. See also *What is Orientation in Thinking?* (*AK* 8:134–135).
41. I will discuss these ideas at greater length in §1.3. As I will argue in §4.1, there is a way of understanding, post-Critically, how ether or Substance can affect the subject even though it is not locomotive.
42. Many commentators make this point, but most overlook the fact that cognition of the alteration of the relatively persistent (e.g., freshly cut wood weathering) requires a concept of substance as much as cognition of the arising/perishing of this relatively enduring thing (e.g., wood being incinerated while ash and smoke arise).
43. Both Sebastian Gardner and Strawson have similar worries. See Sebastian Gardner, *Kant and the Critique of Pure Reason* (London: Routledge, 1999), 179 and Strawson, "Kant on Substance," 270.
44. Here I am referring to Ward's interpretation discussed above.
45. See, for example, Allison, *Kant's Transcendental Idealism,* 229 and Guyer, *Kant and the Claims of Knowledge,* 225. Here one might think of the role that the experiences of objective succession (established in the Second Analogy) and objective simultaneity (established in the Third Analogy) play in Kant's B-edition "proof" for the First Analogy. Experience of objective succession (alteration of substance) and objective simultaneity (causal community of substances), however, requires persistent substance (established in the First Analogy). If one of the arguments is unsound, it will affect the soundness of the others. Unfortunately, with only one concept of substance, at least one of the arguments in the Analogies will be unsound. For example, the argument for the Third Analogy seems to require both Substance (matter everywhere) as well as substances (that stand in causal community).
46. I will return to this idea in the next section.
47. One might object that I am violating my second principle of interpretation insofar as my reconstruction of the Analogies hinges on Kant being inconsistent. As we will see in subsequent chapters, however, my overall interpretation hinges on Kant being able to overcome this inconsistency post-Critically and in such a way that is consistent with his thinking in the Critical period.
48. One of the reasons I think this dilemma has been overlooked in the literature is that commentators tend to translate "*Beharrliche*" as "permanent" rather than "persistent." In *OP,* however, Kant describes the ether (Kant's post-Critical conception of Substance) as "*alldaurend*" which demonstrates that he recognizes the difference between being permanent and being persistent. See *OP* 21:584.
49. Of course, whereas the existence of God would seem to exclude the existence of gods, the existence of Substance does not exclude the existence of substances. Although the items compared in an analogy must be similar in some respects, they need not be similar in every respect.
50. Translation is mine. On the previous page, Kant mentions the 'elementary concepts of the moving forces of matter' (*OP* 21:625.11–12) as being central to the transition, which he elsewhere identifies with the concept of the ether. In *Farrago* 2, Kant likewise claims that the transition is founded on '*a priori* principles' (*OP* 21:639.27) and will 'fill a gap in the system of pure natural science and close the circle of all that belongs to the *a priori* cognition of

nature' (*OP* 21:640.4–6). He further identifies this part of pure natural science with 'the system of the moving forces of matter according to its form' (*OP* 21:640.8–9). Interestingly, in this passage he also distinguishes the "form" of the moving forces of matter (concept's intension) from the "object" (concept's extension). This distinction plays an important role in my reconstruction of Kant's position. Tuschling also recognizes the importance of these passages in identifying the gap in Kant's Critical philosophy. I completely agree with Tuschling that the gap does not lie simply in the *application* of transcendental principles to the objects of outer sense, but rather *within* the transcendental principles themselves. See Tuschling, *Metaphysische und Transzendentale Dynamik,* 182. As discussed in §0.4, however, I do not believe the existence of the gap depends upon any failure of Kant's project in *MFNS*. I am also far more optimistic than Tuschling as to Kant's ability to bridge this gap through his transition project.

51. Insofar as Kant's special metaphysical natural science (*MFNS*) also falls within his pure doctrine of nature, the quote is likewise consistent with the view that the gap is in *MFNS*. Kant's comments in the preface to *MFNS* do not seem to preclude this possibility (see *MFNS* 4:468–470). At the same time, can *MFNS* really be a *pure* doctrine of nature since it relies upon an *empirical* concept (the empirical concept of matter)? In the Introduction to *CPR*, Kant suggests that this is sufficient to make *a priori* cognition *impure* (see *CPR* B3). Much the same could also be said of the dynamical principles (e.g., the Analogies), however, since they rely on 'empirical thinking in an experience' (*CPR* A160/B199). In any case, I have already provided some reasons for why the gap should not be located in *MFNS* (see §0.4). If one accepts my diagnosis of the gap, furthermore, the gap problem would be orthogonal to the concerns of *MFNS*. Kant's discussion in *MFNS* is entirely based upon the concept of matter as something *locomotive*. Since Substance or the ether is not locomotive, it is difficult to see how it is relevant to the "metaphysical doctrine of body" that Kant articulates in *MFNS*. Given this fact, Substance or the ether would seem to be a "transcendental principle" (belonging to the transcendental part of the metaphysics of nature) rather than a "metaphysical principle" (belonging to the special part of the metaphysics of nature). See *CJ* 5:181.
52. Although Rollmann and Hahmann identify what I have called "Substance" in the Analogies with the ether in *OP*, they try to explain Substance/ether using only the category of substance. They recognize that this leaves them with no way of explaining what I have called "substances." See Rollmann and Hahmann, "Weltstoff und absolute Beharrlichkeit," 169, 187, and 189–190.
53. For an excellent discussion of the principle both in *CPR* as well as in *OP*, see Peter Rohs, "Kants Prinzip der durchgängigen Bestimmung alles Seinden," *Kant-Studien* 69, no. 2 (1978): 170–180.
54. For more on illegitimate use of the principle of thoroughgoing determination, see Michelle Grier, *Kant's Doctrine of Transcendental Illusion* (Cambridge: Cambridge University Press, 2001), 245–248 and Rohs, "Kants Prinzip der durchgängigen Bestimmung," 174.
55. See *CPR* A576/B604.
56. I will discuss Kant's theories of affection and perception at greater length in the final chapter.
57. One might claim that this is far closer to the absolute idealism of German Idealism. Either position, however, would be inconsistent with Kant's Critical view.
58. Although I agree with Förster that Kant's conception of the ether bears much in common with the Transcendental Ideal, the former cannot be considered a merely regulative thesis. Lord recognize this fact when she claims that '[the

ether] cannot be merely a regulative idea for our understanding the moving forces as a system; it must also be a constitutive concept for the possibility of outer experience as such.' Lord, *Kant and Spinozism,* 166. In *OP,* the ether is not a hypostatization of the distributive unity of perceptions, but rather a necessary condition for the possibility of the latter. See Lord, *Kant and Spinozism,* 161. This point will come out very clearly, I believe, in my reconstruction of the Ether Deduction.

59. Chapter Three will reconstruct Kant's Ether Deduction. It has been suggested to me, however, that perhaps the Antinomies provide Kant with an alternative argument that possesses constitutive force. The problem is that use of either thesis or antithesis in the Antinomies (e.g., the 'whole of the world' mentioned in the antithesis to the first Antinomy) in order to substantiate Kant's claim for a constitutive totality of empirical reality would seem to commit Kant to a dialectical error. One can even view Kant's warning against hypostatization of the Transcendental Ideal as being a reiteration of the lessons learned in the Antinomies. Since a constitutive function for the totality of empirical reality is not defended in the solution to the Antinomies, they are of no use to Kant in substantiating his claims in the Transcendental Ideal.
60. See also Kant's resolution to the Second Antinomy in *CPR* where he argues that matter is infinitely divisible (*CPR* A523–527/B551–556). In both *MFNS* (4:507) as well as in the resolution to the Second Antinomy, Kant holds that transcendental idealism allows us to hold that matter is infinitely divisible without having to hold that it consists of an infinite number of parts.
61. See *MFNS* 4:505. Friedman emphasizes that, in *MFNS,* Kant adopts a true continuum view where matter is elastic all the way down and is distributed over all the points in space that it fills (i.e., he rejects the discrete force-centers of the pre-Critical *Physical Mondadology*). See Friedman, *Kant's Construction of Nature,* 139 and 185.
62. See *CPR* A169/B211.
63. See *CPR* A170/B212.
64. See *MFNS* 4:503–504. For more on this, see Pollok, *Kants Metaphysische Anfangsgründe,* 267. Pollok argues that the infinite divisibility of matter refers to its *intensive* not its *extensive* quantity.
65. See *MFNS* 4:513. Whereas repulsive force 'fills' space for Kant, attractive force only 'occupies' without filling space. See *MFNS* 4:516 and 535.
66. As we will see in the next chapter (§2.2), the mechanical forces of substances depend upon (without being reducible to) the dynamical forces of the ether or Substance in *OP.* Kant already insists on the dependency of mechanical force on dynamical force, however, in *MFNS.* See *MFNS* 4:536–537 and 551.
67. See *MFNS* 4:563–564. In Reflection 58 from the late 1770s (*AK* 14:460), Kant contrasts *absolutely* empty space (where there is no causal community) from *comparatively* empty space (where there is nothing impenetrable). A space where there was force but no impenetrable matter would be comparatively empty. Although a space filled only with Substance or ether (understood as a penetrable plenum of attractive and repulsive forces) would be comparatively empty, it would not be absolutely empty. For a thorough discussion of Kant's different conceptions of empty space (phoronomical, dynamical, and mechanical) in *MFNS,* see Friedman, *Kant's Construction of Nature,* 512–519.
68. See *MFNS* 4:534.
69. See also Proposition 2 of the Mechanics section of *MFNS,* where Kant connects his First Law of Mechanics with the B-edition principle of the First Analogy. Pollok claims that, in Proposition 2, Kant is unifying his transcendental concept of substance from the First Analogy with his empirical concept of

matter from *MFNS*. Although Pollok quotes *CPR* B227, where Kant equivocates between 'substance or substances,' Pollok does not seem to recognize the complications this might pose for Kant's account in *MFNS*. See Pollok, *Kants Metaphysische Anfangsgründe*, 409.

70. Translation is mine. Although this passage comes from 1798, it is still prior to Kant's attempt to develop an *Elementarsystem* of the moving forces of matter and so prior to his dramatic break from the theory of matter as presented in *MFNS*. For more on this, see §0.3.
71. Hanna adopts an approach along these lines. He thinks of substances as positions or roles within Substance's total system of causal-dynamic forces. Given the fact that mechanically interacting physical bodies (substances) possess many properties that Substance does not possess, however, it is unclear how substances could simply be parts of Substance. Hanna recognizes this fact and insists on strong supervenience without reduction. See Hanna, *Kant, Science, and Human Nature*, 390–408. As I will argue below and in the next chapter (§2.2), I do not think that supervenience is the best way of characterizing the relationship between substances and Substance.
72. In this respect, the solution under consideration returns us to a position quite close to Westphal's interpretation.
73. As mentioned above, the idea that substances exist only *adjectively* on Substance is a popular view in the secondary literature.
74. Similarly, Kant defines "material substance" as 'that in space which is movable in itself, that is, in isolation from everything else existing external to it in space' (*MFNS* 4:503). Although substances fall under Kant's definition of "material substance," Substance does not since it is omnipresent (nothing exists external to it in space).
75. Even in the case of rotational motion around an axis, the rotating body changes its relation to an external space (e.g., the earth turning different sides to the moon every 24 hours). See *MFNS* 4:482. Although Kant will describe the ether as 'internally moving,' this motion is never in relation to an external space.
76. See *MFNS* 4:483.
77. As we will see in the next two chapters, the *locomotion* of substances depends upon the *internally* moving forces of the ether or Substance in *OP*.
78. Kant argues in later sections of *OP* that the ether can affect subjects through its internally moving forces. I will discuss this issue at greater length in §3.1.
79. A "body" just is 'a matter between determinate boundaries (which therefore has a figure)' (*MFNS* 4:525). Insofar as *matter* is essentially locomotive, only substances could count as matter. Insofar as bodies (as a determination of matter) have determinate boundaries, only substances (spatially discrete) could count as bodies.
80. In choosing the term "emergence" over "supervenience," I want to emphasize the fact that substances are something over and above Substance. Unlike David Lewis' conception of supervenient entities, substances are not "ontologically innocent" and possess their own causal efficacy. See David Lewis, *Parts of Classes* (Cambridge: Blackwell, 1991), 81. One might say that the dynamic activity of Substance is *necessary* though not *sufficient* for the causal efficacy of substances. I believe this is important for the success of the Third Analogy for the reasons mentioned above. As we will see in the final chapter, substances ultimately emerge from two sources: (1) the physical activity of the ether, and (2) the mental activity of the subject. For a contemporary view on dynamic emergence, see Timothy O'Connor and Hong Yu Wong, "The Metaphysics of Emergence," *Noûs* 39, (2005): 658–678.

81. Although limited to the development of organisms, one might argue that the theory of epigenesis Kant endorses is a form of emergence. See *CJ* 5:423–424.
82. Insofar as one grants that Kant's conception of Substance is continuous with his conception of the totality of empirical reality in the Transcendental Ideal, it is important to note that Kant holds that the totality of empirical reality, understood as the 'original being,' is not 'a mere aggregate of derivative beings' (*CPR* A579/B607). Kant rejects the part-whole relation in favor of a ground-consequence relation.

2 The Development of Kant's Ether Theory

> There exists a matter, distributed in the whole of space as a continuum, penetrating all bodies uniformly (so not subject to displacement) which may be called the ether, the caloric, or whatever, but which is no hypothetical material (in order to explain certain phenomena, and more or less ostensibly imagining the causes for given effects) but can rather be recognized and postulated *a priori* as a part belonging necessarily to the transition from the metaphysical foundations of natural science to physics . . . It is to be acknowledged as a primordially moving material—not hypothetically invented, but one whose forces give it reality and which underlies all motion of matter; a continuum which, taken in its own right, forms a whole of moving forces, whose existence is known *a priori*.
>
> —Kant, from *Übergang* 2 (*OP* 21:218 and 223–224)[1]

2.0 INTRODUCTION

The above quote summarizes much of Kant's mature view on the ether. As mentioned in the last chapter (§1.1), Kant's conception of Substance in the Analogies from *CPR* is the conception of a unified material whole that is both sempiternal and omnipresent. It is clear that Kant believes that the ether possesses these properties as well. In a fragment from *Convoluts* 10–11 of *OP* discovered after the publication of *Kants gesammelte Schriften* and written immediately after the *Übergang* section, Kant describes the ether in terms of its 'omnipresence [*allgegenwart*] and continual persistence [*continuirliche Beharrlichkeit*].'[2] While Kant's post-Critical conception of the ether certainly encompasses both of these properties, it goes far beyond them as well. At the same time, his Critical conception of the ether could not be further removed from his post-Critical conception of it. As mentioned in the Introduction, when Kant affirms the actuality of the ether, he is neither simply affirming those conceptions of the ether popular in his day (light-ether/heat-ether), nor is he affirming the reality of the static medium that Michelson and Morley failed to detect in the 19th century. Some commentators place too much emphasis on the *name* of this material to the exclusion of its *function*. As the above quote makes clear, the name that one calls this material whole is not important, but what is important is the material whole's function, which depends on it possessing certain properties. Its dynamic activity is the ultimate source of perceptual affection and the material ground for physical bodies that subjects experience in space and time.[3] Mechanical forces (locomotion in space) depend upon these dynamic forces (internally moving).[4] Since it is omnipresent and sempiternal, it precludes the experience of empty space and time.[5] Kant believes the ether is a collectively moving unified material whole, continuously expanded, and a constantly agitating plenum of dynamic (attractive and repulsive) forces. As one can see in the above quote, matter is itself realized through these forces. One could metaphysically summarize many of these properties by saying that the ether is a compositionally plastic, intrinsically structural substrate of dynamic forces. In more Kantian terms, the ether is the systematic unity of the moving forces of matter.

The purpose of this chapter is to examine how Kant's conception of the ether develops between the Critical and post-Critical periods so as to solve certain important problems from the Critical era. Although all of the problems I discuss in this chapter are on the periphery of the gap I located in Kant's Critical philosophy, they are important to understanding the immense role that the ether comes to play in Kant's post-Critical philosophy, and the way in which Kant ultimately fills the gap (see Chapter Four) will be instrumental in solving some of these problems. Although my primary focus in other chapters is on how *OP* solves certain problems from *CPR* (the most important of which is the gap problem), this chapter will identify some problems with Kant's other Critical-era works (specifically *MFNS* and *CJ*) that Kant's post-Critical philosophy can also be seen as solving.

This chapter is broken into three sections. The first section will examine Kant's conception of the ether in the Critical period as well as why he felt compelled to develop a dynamic theory of matter. This section will also discuss some important problems from *MFNS* that I first touched on in the Introduction (§0.3). The second section will evaluate whether Kant's new ether theory, developed in the early sections of *OP*, will allow him to solve these problems. The final section will argue that post-Critical developments in Kant's ether theory help him to explain the unity of nature in a way that he could not using his Critical conception of the ether in *CJ*.

2.1 KANT'S CRITICAL CONCEPTION OF THE ETHER

As I argued in the last chapter, one can view Substance in the Analogies of *CPR* as equivalent to the "totality of empirical reality" Kant discusses in the Transcendental Ideal. There Kant also refers to the totality as the "*omnitudo realitatis*" (all of reality).[6] This is a concept that shows up frequently in the pre-Critical writings and Kant, at several points, tries to prove the existence of the *omnitudo realitatis* which he identifies with God.[7] Likewise, in the Transcendental Ideal, Kant identifies the idea of the *omnitudo realitatis* with the idea of God, but he breaks from the pre-Critical writings insofar as he holds that there is no proof for the *existence* of such a being.[8] If I am right, however, then divine or not, Kant requires an argument for the existence of the totality of empirical reality, which is merely presupposed in the Transcendental Ideal. As I will demonstrate in the next chapter (§3.3), Kant offers such an argument in the Ether Deduction of *OP*.

When it comes to the ether specifically, Kant makes only one oblique reference to it in *CPR* and not in a flattering light.[9] In the Postulates of Empirical Thought, Kant considers the possibility of something persistently present in space without at the same time filling that space an 'intermediate thing' that might occupy some ontological position between matter and mind. Unlike the categories, however, the possibility of this material cannot be ascertained *a priori* as a condition for experience, but rather must be a concept given *a posteriori* through experience. Since there is no experience of this material, however, the concept of it is nothing but a 'figment of the brain' and 'entirely groundless' (*CPR* A222–223/B269–270). As the quote that begins this chapter makes clear, Kant views the ether, in *OP*, as something that both persists and completely fills space (like Substance in *CPR*). Contrary to his claims in *CPR*, however, Kant comes to view the ether no longer as a mere thought entity or figment of the brain, but rather as something whose existence can be established *a priori*. The reasons for this change of heart will be made clear below.

Just as in *CPR*, in *CJ* Kant mentions the ether only in passing. At one point, he uses the ether to illustrate what he means by a 'matter of opinion.' He describes it as an 'elastic fluid penetrating all other materials' (*CJ* 5:467) and suggests that it is the kind of thing that is in perceivable in principle though not in practice. As I will argue in the next chapter, however, Kant comes to reject the idea that the ether is even in principle perceivable (as substances are) in the *Übergang* section. In the third edition of *CJ*, Kant seems to adopt Euler's view that colors are due to the uniform vibration of light-ether.[10] At another point, Kant describes the caloric (heat-ether) as a fluid whose accumulation or loss can explain the aggregations of matter (gas, liquid, solid).[11] Notwithstanding Kant's apparent acceptance of both light-ether and heat-ether in *CJ*, his post-Critical conception of the ether as a plenum of attractive and repulsive force cannot be reduced to the conceptions of either light-ether or heat-ether common in his day.[12] Even though a fully dynamic conception of the ether is absent from Kant's discussion in *CJ*, it will nevertheless come to play an important post-Critical role relative to *CJ*.

Whereas the ether plays a minimal role in *CPR* and *CJ*, the ether comes to play an important, albeit regulative, role in the General Remark on the Dynamics from *MFNS*.[13] It allows one to think of all space as full and so can 'controvert an [atomist] hypothesis (of empty spaces)' (*MFNS* 4:534). In this context, Kant views the ether as the limiting case for his theory of density. Whereas each *kind* of empirical material has a specific repulsive force, the attractive force is proportional to the *amount* of material in question and accounts for the density of the object in question.[14] The ether is something which, given its lack of quantity, has an incomparably greater repulsive force than attractive force in comparison to any other material. Thus, the concept of the ether would allow one to think of a lack of quantity within a given volume without appeal to empty spaces.

Later on, in the Phenomenology chapter of *MFNS,* Kant uses the ether again though still only regulatively. This time he uses it to account for cohesion, which he understands as the external compression of a given body by the ether.[15] The ether applies its cohesive force by virtue of the universal gravitation which operates throughout it. If one assumes that the ether occupies space external to bodies, the space inside of bodies must be full as well since empty space within a body could not resist the repulsive force the ether exerts on the material which constitutes the body. Consequently, empty space, 'although not logically impossible, would still be so dynamically, and thus physically' (*MFNS* 4:564) either inside or outside a body.[16] Given its lack of quantity, however, the ether does not provide mechanical resistance to the bodies that move through it (e.g., the motion of heavenly bodies).

Even though the ether seems to perform several important functions within the context of *MFNS,* the ether itself remains merely a *hypothesis,* though one that is preferable to the atomist's reliance on empty space. It is clear that Kant's *bête noir* is atomism, but why is he so resistant to the atomist position? Some of Kant's problems with atomism should already be clear from the foregoing discussion of the Analogies. Whereas the atomist makes use of empty spaces separating atoms, Kant does not consider such empty spaces to be possible objects of experience. As he says in the Third Analogy, if one were to perceive empty space, experience would be "broken off." Secondly, Kant sees the atomist position as requiring concepts like absolute impenetrability which is a mere *qualitas occulta.*[17] The atomist must rely on something like the principle of contradiction to explain the impenetrability of atoms, i.e., the reason why an atom is impenetrable is because its impenetrability is contained within the concept of "atom." In a rare example of Kant's humor, however, he quips that the 'principle of noncontradiction does not repel' (*MFNS* 4:498). According to Kant, the atomist's conception of impenetrability is an 'empty concept' and gives up 'all forces *inherent* in matter' (*MFNS* 4:525). For Kant, impenetrability requires *physical* force and for any physical force, a greater can be imagined to overcome it. Consequently, there can be no such thing as *absolute* impenetrability.[18] For much the same reason, Kant rejects the concept of physical simples.

Additionally, if one accepts transcendental idealism and Kant's argument in the Second Antinomy, any claim to physical simplicity is itself dialectical, i.e., an example of reason trying to know that which cannot be given in any possible experience. If objects are merely appearances, and if any space an object occupies is itself mathematically divisible, then the appearance which occupies that space is divisible as well. Whereas atoms are defined in terms of their intrinsic non-relational properties (i.e., properties they possess regardless of their relationship to other things), objects as they appear possess intrinsic relational (i.e., spatiotemporal) properties that cannot be captured by an atomistic model. Although appearance is divisible *ad infinitum,* this does not entail an *actual infinity* of parts which is no more an object of possible experience than physical simples are.[19] As we will see below (§5.3), Kant's rejection of physical simplicity goes hand-in-hand with his rejection that matter can be mind-like or representational. Kant retains his basic reasons for rejecting atomism in the post-Critical period and, as I will show below, his rejection of atomism plays an important role in the Ether Deduction.

In the early sections of *OP,* Kant continues his attack on atomism, claiming that it has 'no place in the philosophy of nature' (*OP* 22:207). His objections to atomism are familiar from *MFNS.* Kant continues to insist upon the infinite divisibility of matter, consequently rejecting the atomist's reliance on physical simples.[20] Kant also objects to the atomist's reliance upon empty space, since it is no object of possible experience.[21] These criticisms of atomism are ones that Kant will continue to repeat throughout the entirety of *OP*, including the Ether Deduction itself.[22]

The atomist also faces a serious problem with cohesion, via corpuscularianism, which Kant takes to be simply a species of atomism.[23] Corpuscles are physical simples that possess their own size, shape, and motion non-relationally. Corpuscularianism is a form of atomism since it asserts that corpuscles are physical simples which are both absolutely impenetrable and indivisible while also separated by empty space. Corpuscles come together to form physical objects according to their respective geometrical shapes and sizes which are conducive to their fitting together to form complex objects.[24] As an analogy, one might think of the pieces of a jigsaw puzzle fitting together to form the overall picture.

As mentioned above, although Kant naturally objects to the corpuscularian's reliance on empty space, there is another problem with the corpuscularian view that deserves mention as well, since it is specific to the issue of cohesion. The problem is that nothing beyond the concept of indivisibility accounts for the cohesion of the mathematical parts (e.g., the left and right hand sides) of the constitutive corpuscles. Yet unless the corpuscularian can explain the cohesion of corpuscles, she cannot ultimately explain the cohesion of the objects that those corpuscles form. At this point, the corpuscularian can only appeal to the absolute indivisibility of corpuscles, i.e., their lack of proper parts, in order to explain the cohesion of a corpuscle's mathematical parts. As mentioned above, however, Kant rejects such explanations as vacuous. To see the problem, one might think of the doctoral examination scene in Moliere's *The Imaginary Invalid* where a doctor asks a candidate why opium makes one sleep and is told by the candidate that opium makes one sleep because of its 'dormitive virtue' [*virtus dormitiva*]. Although the examination committee was happy with the candidate's answer, Kant surely would not have been.

In the early sections of *OP*, Kant uses the ether to explain not only the same physical phenomena that the atomist has trouble explaining, but also physical phenomena that his own theory of matter in the General Remark on the Dynamics of *MFNS* seems incapable of explaining. Although there are many examples, the most important is Kant's discussion of cohesion in *OP* which has implications for density as well. As noted in the Introduction (§0.3), Kant's explanation of density in *MFNS* seems to be circular. Of course, the atomist is no better off since she must appeal to empty spaces between atoms in order to account for the different densities of physical objects. If one accepts Kant's criticisms of atomism, however, such an appeal to empty spaces would seem to be a non-starter.

Westphal argues, surprisingly, that Kant's own theory of matter in *MFNS* commits him to a corpuscular explanation of density as well. As mentioned above (§0.3), the volume of a given matter is determined by the mutual limitation of attractive and repulsive forces. Attractive force is constant across all materials. Repulsive forces vary by material, but are only surface forces that account for impenetrability. Density is determined by the intensity by which the attractive/repulsive forces of matter fill a given space. Since attractive and repulsive forces operate in all directions from a central point, different materials will form into different sized spheres depending upon the strength of the repulsive force specific to the material in question. The boundary of the sphere is that point at which the scalar intensities of the attractive and repulsive forces are equal. Different materials *must* differ in volume and for the same reason different materials *cannot* differ in density since the volume of the sphere will always be proportional to the intensity of the repulsive force. The only way that Kant can account for differences in density would be on the basis of the number of vacant interstices that different material spheres form when combined into composite bodies. Consequently, it does not seem to be the case that Kant's theory of density is any better off than the corpuscularian view.[25]

Even assuming that Westphal is right about Kant's view, it is important to remember Kant's problem with vacant interstices. The problem is not the interstices per se, but rather the fact that such interstices are vacant. Kant rules out explanations involving vacant interstices because he rules out empty space as an object of possible experience, which is just what the vacant interstices view would seem to require. If one assumes the existence of omnipresent ether, however, would these interstices be vacant after all? One might object that an ether theory might just as easily be added on to a corpuscularian theory in order to avoid Kant's objection to the experience of empty space. Even if one did this, however, it still would not overcome all the other problems that a corpuscularian theory faces, e.g., questions of impenetrability, indivisibility, and cohesion. When it comes to impenetrability, Kant's dynamic theory has a distinct advantage over the corpuscularian view. Whereas absolute impenetrability is an unexplained occult quality for the corpuscularian, Kant can explain the relative impenetrability of a material sphere on the basis of its repulsive forces.[26] Unlike the corpuscularian, Kant does not commit himself to indivisible simples and so does not have to explain how indivisibility is possible. Even if Kant's dynamic theory is superior to the corpuscularian view, can it overcome its own internal problems? This will require us to examine how Kant's ether theory develops in the early leaves of *OP*.

2.2 KANT'S POST-CRITICAL CONCEPTION OF THE ETHER AND THE *METAPHYSICAL FOUNDATIONS OF NATURAL SCIENCE*

When it comes to cohesion, in the early leaves of *OP*, Kant claims that the ether is perpetually oscillating and that this causes the parts of matter to coalesce into determinate bodies. Much as matter in general is constituted by both original attractive and repulsive forces in *MFNS*, so too is the ether in *OP*. If the ether had only original attractive forces, it would collapse into a point. If the ether had only original repulsive forces, it would expand until it dissipated into empty space. The difference between the ether in *OP* and matter in the Dynamics of *MFNS* is that the original attractive and repulsive forces are *alternating*, which causes a perpetual *oscillation*.[27] As *perpetually* oscillating, these *internally* moving forces produce *external* motion in the world and prevent it from reverting to a 'state of lifeless stasis' (*OP* 21:310).[28] In addition, the *oscillation* forces the parts of matter to cohere and determines the specific densities of the resulting bodies. The *contact* attraction of cohesion is independent from the quantity of matter (or density in a given volume) and so distinct from the *penetrating* original attraction of gravitation. Besides offering a new account of cohesion, is Kant also offering a solution to the circularity problem that faces his theory of density in *MFNS*?[29]

In my opinion, Kant's post-Critical ether theory does not offer him an escape from the circularity problem.[30] Assuming that original attraction = gravitation, if gravitational force is proportional to density in a given volume, and the density of an object depends upon the degree to which the ether is oscillating, and the oscillation of the ether assumes the interplay of original attractive and repulsive forces, then doesn't gravitational force assume gravitational force or original attraction? The solution might be to say that universal gravitation is not identical to original attraction.[31] Whereas the former would be proportional to density in a given volume, the latter would not. Such a solution, however, seems unsupported by the text either in the Critical or the post-Critical periods. Both in *MFNS* as well as in the letter to Beck, Kant identifies original attraction with universal gravitation.[32] This identification continues in *OP* when Kant offers the above account of density based on the idea of oscillating ether.[33]

Notwithstanding Kant's failure to overcome the circularity problem, what reason do we have for believing that matter is constituted by the original forces of attraction and repulsion whether oscillating or not?[34] We have already seen (§0.3) that Kant's stated argument for original repulsion and attraction in the Dynamics chapter of *MFNS* rests on a non-sequitur, *viz.* the move from impenetrability to repulsive force (where the latter then requires attractive force to limit it). What if Kant were to infer the two basic moving forces from the two directions of motion (approach or recede), the only alterations of spatial relations that Kant's conception of space allows?[35] It is only if one assumes that the concept of matter can really be *constructed* from space and time as *a priori* forms of intuition in conjunction with the categories and Principles of Pure Understanding, that such an inference makes sense. Westphal argues that although Kant might be able to use this approach to establish that there are only two different *kinds* of force, he cannot establish that there are only two basic *forces*. As genera, the former could not be constitutive explanations but would rather only be regulative classifications.[36] Konstantin Pollok notes that Kant's dynamic concept of matter cannot be constructed *a priori* in intuition since the foundational forces are intensive magnitudes (qualities) that must be sensed *a posteriori*. If Kant's concept of matter cannot be constructed *a priori,* however, then his doctrine of nature in *MFNS,* based on this concept, would fail to be scientific.[37]

Perhaps there is another way of establishing the existence of the two basic forces. Assuming that the mechanical (atomistic or corpuscularian) account of matter is ruled out for the reasons mentioned above, there seems to be a way of arguing for two basic dynamic forces counterfactually—that is, based on what the world would be like if, contrary to fact, there were only one or the other force. More precisely, Kant imagines what the world would be like with only basic attraction (a physical point), and then what the world would be like with only basic repulsion (extended to infinity). Finding bodies that are neither physical points nor infinitely extended, Kant can conclude that matter must have both basic attractive and repulsive forces. It does not seem that Kant's derivation of the two basic forces hinges simply on the phoronomic character of matter. Secondly, and more importantly, even if there is only one *basic* attractive force, this does not seem to preclude other *derivative* attractive forces. Kant certainly thinks this is the case with cohesion and the same might also be true of universal gravitation even if Kant himself did not make this move.

Returning to the idea that cohesion requires the oscillation of the ether, one might also wonder why solid bodies cannot change their shape as easily as fluid bodies given that both kinds of body cohere because of the oscillations of the ether. Kant's explanation is to say that bodies solidify due to the *escape* of ether, whereas the fluidity of bodies is due to the *penetration* of ether.[38] In both respects, the ether is itself treated like a fluid material. This raises a further problem. If the fluidity of a body is due to the ether and the ether is itself fluid, then what accounts for the fluidity of the ether?[39] Realizing that this could lead to an infinite regress of explanations, Kant claims that the ether is not itself fluid, but is rather that which accounts for the fluidity of a given body. Bodies, whether solid or fluid, have certain positive characteristics. Bodies are ponderable since they are capable of being weighed to determine the quantity of matter they contain. Bodies are coercible since they can be forced into certain forms or bounded in certain ways. Bodies are formed through the cohesion of their parts. Bodies possess a certain amount of matter that can be exhausted. Kant sums up his position on the formation of rigid and fluid bodies in the A *Elementarsystem* 7:

> There must be a material that cannot be thought of as rigid, but rather as radical fluidity, assumed as the first cause of all other estimable fluidity, that is laid as the ground within which inner restless vibrations spread the heterogeneous parts of each ponderable material to build rigid bodies of determined texture and figure, and so move fluidity into the state of rigidity, namely through moving force which causes the fluid, merely through the heterogeneity of one or another thoroughgoing materials, to form into fibers (rays), sheets, and blocks (in the three dimensions of space) according to the difference of their specific ponderability in which each primitive fluidity diminishes the inner motion of exhausted materials and so the materials are shaped into small differentiated parts.
>
> (*OP* 22:594.13–25)[40]

Kant's position is that the oscillations of the ether (moving forces) push the parts of matter into positions where the motion of the ether is most diminished. There matter coalesces into bodies according to the three dimensions of space. The rigidity of the body will be inversely proportional to the activity of the ether, whereas the fluidity of a body is directly proportional to the activity of the ether. The lower the frequency at which the ether oscillates, the more rigid the body becomes. The higher the frequency at which the ether oscillates, the more fluid the body becomes.[41]

Jeffrey Edwards and Martin Schönfeld argue that Kant's theory of body formation anticipates contemporary field theory. They see the ether as an 'energetic continuum, or a spatial field of natural forces.'[42] Under this view, physical particulars are understood as focal points from which forces are exerted. These physical particulars are not causally self-sufficient, however, and receive whatever forces they possess from the field upon which they supervene.[43] Robert Hanna has also proposed something very much like a field theory interpretation of the ether where physical particulars are nothing but spatiotemporally limited positions within this total field of attractive and repulsive forces. In other words, they are parts of the whole of the ether.[44]

Although I agree that the ether is essentially a field of attractive and repulsive forces for Kant, as mentioned in the last chapter (§1.3), I prefer an emergence account of the relationship between physical bodies and the field. Physical bodies emerge from the attractive and repulsive forces within this field without being reducible to (i.e., simply parts of) the field itself. At the same time, physical bodies do not supervene on the field insofar as this is understood in terms of the bodies being ontologically "innocent," i.e., entities that only exist adjectively and are nothing over and above the field itself. One should not understand physical bodies, furthermore, as ontologically "guilty" entities that supervene upon the field but are causally impotent (in philosophy of mind, one would say "epiphenomenal"). An emergence account allows physical bodies to exist as entities over and above the field (ontologically guilty) while maintaining their own causal efficacy. This ontology also offers Kant a way of defending himself from Westphal's criticisms of the metaphysical causal thesis from *MFNS* (discussed in §1.1). Physical bodies would not only be distinct from one another (spatially discrete), but also distinct from the field from which they emerge. Every change in a body would have an *external* cause understood proximately in terms of the motion of another body or distally in terms of the dynamic forces of the field upon which the mechanical forces ultimately depend.[45]

As I will discuss at greater length below, although the attractive and repulsive forces of the field are *necessary* for the existence of physical bodies, their properties, and relations, the field is not *sufficient* for any of these things. In particular, as I will argue in Chapter Five, one cannot ignore the role the subject plays in constituting physical bodies (including itself as an embodied cognitive subject). At this point, however, it is most important to note that Kant's ontology of matter is one of *dynamic forces* not of *compositional stuff* and all physical bodies depend upon (in the sense of a necessary condition) the original field of attractive and repulsive forces for their existence and for their determinate properties. Kant's subscription to a basic ontology of force marks a distinct break with the mechanistic atomists that precede him. Although the above quote from the A *Elementarsystem* is unclear as to whether moving force or compositional stuff is ontologically fundamental, as the quote that opens this chapter makes clear, Kant ultimately views moving force as the fundamental aspect of his ontology.[46]

This conception of the ether is also relevant when it comes to determining the material properties of bodies. As I have repeatedly noted, in the General Remark on the Dynamics in *MFNS*, Kant warns against attempting to determine the material properties of bodies *a priori*.[47] In *OP*, however, Kant sees the need for determining these material properties *a priori*. Although he speculates on these issues in *MFNS*, Kant is unable to account for phenomena like cohesion, density, and ponderability using only the categorically determined empirical concept of matter. Kant recognized that any account of these phenomena would require a conception of matter more specific than the one that *MFNS* provides. According to Kant, the ether is going to be the material condition of properties like cohesion, density, and ponderability. In a fragment from *Convoluts* 10–11, discovered after the publication of *Kants gesammelte Schriften,* Kant says that the whole of moving forces is 'not ponderable etc., but rather causes these properties in the movable.'[48]

In the *Elementarsystem,* Kant sees the ether as something which makes all of these properties of bodies possible, without itself possessing any of these properties. In the case of ponderability, and as mentioned above, Kant characterizes the ether as 'radical fluidity' (*OP* 22:588) which is not itself ponderable, but that which makes ponderable fluids possible.[49] Kant recognizes that an *absolutely* imponderable material is contradictory since it would be matter without quantity, but he insists that matter can still be *relatively* imponderable insofar as it pervades ponderable bodies.[50] Kant goes on to give a purely negative characterization of the ether in terms of the properties that physical bodies (or what I have called "substances") might possess: 'Imponderable—Incoercible—Incohesible—Inexhaustible. That all of these moving forces stand under the system of categories and that one universal [matter] primitively underlies them all' (*OP* 21:183).[51] In other words, the ether cannot be weighed, forced, cohered, or exhausted. Even so, the ether is what makes the weighability, forcing, cohesion, and exhaustion of physical bodies (either fluid or solid) possible.[52] I would suggest that Kant had in mind the *Elementarsystem* (which he was about to begin) when he refers to a 'schematism of the concepts of metaphysics' (*OP* 21:169).[53] Although the ether or Substance does not possess any of the categorical properties that physical bodies or substances possess, it does still possess categorical properties, but these are properties that bodies or substances cannot possess. Later, in the *Übergang* section, Kant says that the ether is universally distributed in space (quantity), all-penetrating (quality), internally moving (relation), and physically necessary (modality).[54] When reflecting on the view he has presented, Kant says:

> In all, these theories are in no way grounded upon experience and they borrow nothing from physics, but rather proceed merely from concepts of the possibility of certain effective causes according to laws of motion insofar as they make experience possible *a priori* and as they are necessary assumptions for experimentation.
>
> (*OP* 22:595.14–18)[55]

As mentioned above (§0.3), this marks a radical departure from Kant's project in *MFNS.* Whereas in *MFNS* Kant takes an empirical concept of matter (as the movable in space) and then goes on to determine this concept in accordance with the categories, he now takes an *a priori* concept of matter (internally moving forces) and then goes on to determine this concept in accordance with the categories to generate an *Elementarsystem* of the moving forces of matter. Immediately before introducing the concept of an *Elementarsystem* for the first time, Kant sketches his project: 'Quantity ponderable or imponderable—Quality coercible——incoercible—Relation cohesible (*coalescibel*)——incohesible (*incoaelesc*)—Modality exhaustible——inexhaustible' (*OP* 21:531.5–9).[56] In accordance with the categories of quantity, the moving forces of matter can be either ponderable or imponderable. Kant considers the act of weighing simply as the way one determines the quantity of matter that a given object possesses.[57] Consequently, it is quite easy to understand why Kant places the ponderable/imponderable dyad under the category of quantity. Why, however, does Kant characterize the ether as either ponderable or imponderable? Although the ether is not itself ponderable, it is what makes the ponderability of bodies possible. As mentioned above, the ether is only relatively though not absolutely imponderable since the latter would involve a contradiction. The easiest way to think about it is that the ether is imponderable considered in itself, but ponderable considered in its effects, i.e., the object that is weighed, as well as the instrument of weighing. As Kant says:

> Thus the ponderability of matter is not a property knowable *a priori* according to the mere concept of the quantity of matter; it is, rather, physically conditioned and requires the presupposition of an *internally* moving matter which results in the immobility of the parts in contact with one another [in the lever-arm], by itself being mobile inside this matter. We know of no other matter to which we have cause to attribute such a property, except caloric. Thus, even ponderability (represented subjectively as the experiment of weighing) will require the assumption of a matter which is not ponderable (*imponderabilis*); for, otherwise, the condition for ponderability would be extended to infinity, and thus lack a foundation.
>
> (*OP* 22:138)

The same point goes for the other categories. Although Kant constantly struggles with explaining how these various property dyads fall under their respective categories, the general point remains the same.[58] For any given set of categories, the positive/negative property dyad that falls under the set of categories in question reflects the idea that the ether can only be negatively characterized in itself relative to these properties, but can be positively characterized relative to these properties through its effects in the phenomenal world.[59] Although the ether (or Substance) is a sempiternal and omnipresent plenum of attractive and repulsive forces, it does not possess any of the properties that physical bodies (or substances) possess. As will become clear later on, this negative conception of the ether is important in understanding why the ether, unlike the objects which emerge from it, is not itself an object of experience. What should be abundantly clear at this point, however, is that the relationship between physical bodies and the ether is not simply the part-whole relationship (a view I criticized in §1.3). For example, whereas a whole of water and any of its macrophysical parts share the same basic categorical determinations (ponderable, coercible, cohesible, exhaustible), the ether shares none of these categorical determinations though, if we accept Kant's account, it is what makes these categorical determinations possible in physical bodies. In other words, these physical bodies *emerge* from, but are not *reducible* (in terms of the part-whole relation) to the ether.

Although much of what Kant says in the *Elementarsystem* and other early sections in *OP* is certainly interesting from a historical standpoint and also provides a powerful foil to the atomist's position, it is more an example of natural *philosophy* than it is of natural *science,* since Kant's claims rely neither on experimentation nor on careful mathematical formulation. At the same time, it is important to note that many of the issues that Kant tackles in the early leaves of *OP* (e.g., cohesion, density, body formation) stem from problems he acknowledges with his account in the General Remark on the Dynamics of *MFNS.* As I have argued (§0.4), I do not believe that these problems, severe as they might be, can constitute a gap within Kant's Critical philosophy insofar as they are limited to the completion of Kant's *doctrinal* task (i.e., completing the metaphysics of nature) which he separates from the Critical philosophy itself in the Preface to *CJ.* In other words, Kant's desire to *apply* the Critical philosophy to the domain of empirical physics does not constitute a gap *within* the Critical philosophy itself.

Even so, one cannot ignore the ever more important role the ether comes to play even within these early sections of *OP.* Whereas the ether plays only a regulative role within the framework of *MFNS,* it comes to play a constitutive role in *OP.* It does no good to say at this point that Kant's solution proceeds only upon the regulative assumption of the ether, or proceeds only "as if" the ether existed.[60] This idea is just as problematic as the view discussed in the previous chapter that takes the Transcendental Ideal as a mere problematic concept. Although it is easy enough for me to act *as if* I am free, on the contrary, assuming Kant's view in *OP,* it would be *impossible* for bodies to cohere and for them to possess the physical properties they possess if the ether did not exist. Granting that bodies do cohere and that they possess the properties that they appear to possess, then the existence of the ether is presupposed constitutively.[61]

2.3 KANT'S POST-CRITICAL CONCEPTION OF THE ETHER AND THE *CRITIQUE OF THE POWER OF JUDGMENT*

Although I have spent some time explaining how the post-Critical development of Kant's ether theory relates to certain problems from *MFNS,* in this section, I will examine the relationship between this theory and an important problem from *CJ.* In *CJ,* Kant claims that mechanical laws (i.e., those that describe matter in motion) underdetermine the kinds of characteristics we see in organisms. As he famously says with regard to mechanical explanations of organisms, 'it would be absurd for humans even to make such an attempt or to hope that there may yet arise a Newton who could make comprehensible even the generation of a blade of grass' (*CJ* 5:400). Unlike inorganic beings, organisms generate others of their same kind, themselves as individuals, and the parts of organisms exist for the sake of one another.[62] Although artifacts can also share the last characteristic (e.g., the parts of the watch exist for the sake of one another insofar as together they allow the watch to tell the time), they do not share the other two (watches do not beget watches and they do not generate themselves). Even so, the analogy with artifacts is useful insofar as the latter are designed according to a certain *purpose* (e.g., the watch is designed to tell the time).[63] Kant gives the example of a bird whose hollow bones, wings, and tail all seem to exist for the sake of one another insofar as they are all aimed at a common purpose, *viz.* flight. Inorganic matter in accordance with natural law 'could have formed itself in a thousand ways without hitting precisely upon the unity in accordance with such a rule' (*CJ* 5:360). No analysis of inorganic matter in accordance with mechanical laws could explain why the bird is put together the way that it is, nor could it explain why an artifact is put together the way that it is. Unlike an artifact whose productive cause is external to it (the designer), however, an organism's productive cause is internal to it (which allows it to be self-generating and propagating).[64] This allows us to understand the organism as a natural purpose.[65] Although reflective judgment's regulative assumption of purposiveness drives our *investigation* of organic nature, it cannot constitutively *explain* organic nature since this would be a function of determining judgment.[66]

According to Paul Guyer, this leaves Kant in a difficult position. Although the thesis is only regulative, in *CPR* Kant claims that nature needs to be brought under a single kind of causation or 'fundamental power' (*CPR* A649/B677) if it is to be systematic. As we will see, establishing *constitutively* that nature is a system will be an important goal of Kant's post-Critical transition project.[67] The worry is that Kant may have empirically confuted the single power thesis (irrespective of its regulative or constitutive status) in *CJ* by claiming that *efficient* causation in accordance with mechanical laws is, in fact, insufficient to explain all of nature.[68] If organic nature cannot be understood without the assumption of *final* causation or purposiveness, it would seem as if Kant is committed to two fundamentally different systems of nature (inorganic and organic) that cannot be unified with one another.

It is important to note that, in *CJ*, Kant claims that an organism is not a 'mere machine' (*CJ* 5:374) since it contains not only a motive power but also a formative one. As Kant argues in the *Elementarsystem* of *OP*, however, the ether also possesses a formative power through its internally moving forces. Near the end of the *Elementarsytem*, this formative power allows Kant to claim that organisms are 'natural machines' and that '*final causes* belong equally to the moving forces of nature' (*OP* 21:184–186). Kant suggests that we should understand this final causation in analogy with the way we produce motion in our own bodies as natural machines, *viz.* as the 'willing of an effective cause' (*OP* 21:210). This is not to say that the ether has intentionality, but rather, that we can *think* of its self-moving (or formative) power as analogous to our own intentionality.[69] Later, in *Convolut* 7, Kant again suggests that organic bodies can arise from the ether:

> Life, however, stems from a distinct substance, from an *archeus* (animated matter is contradictory), and organic bodies stand, through the ether, in the relation of a higher organ toward one another.
>
> (*OP* 22:421)

One could read this passage as suggesting that the ether is the *archeus* that implicitly contains all of the functional and formative principles of organisms. As a 'distinct substance,' the ether could be considered a genuine *omnitudo realitatis* presupposed not only for the determination of inorganic bodies but for organic bodies as well. Most importantly, if *all* bodies (whether organic or inorganic) emerge from the ether, Kant would be able to unify both inorganic as well as organic nature within a single world system.[70] As we will see in Chapter Four (§4.4), this is vitally important for the completion of Kant's transition project which requires establishing the systematic unity of nature. At the same time, however, I will argue in the final chapter (§5.5) that embodied cognitive subjects (e.g., rational human beings) possess *mental* powers that do not emerge from the *physical* powers of the ether notwithstanding the fact that both inorganic and organic bodies emerge from the physical powers of the ether.[71] Although this will not confute the single power thesis when it comes to Kant's explanation of *nature,* it does imply that not all powers belong to nature.

* * *

In what follows, I hope not to rely on the specific positions Kant adopts in empirical physics to show how his post-Critical project can fill the gap in his Critical philosophy. Although I will continue to rely on Kant's *general* theory of material nature as developed in *OP,* I will avoid relying on his explanations of *specific* physical phenomena like cohesion and density. Admittedly, in *MFNS* and early sections of the *OP,* Kant seems to suggest that the ether possesses certain physical properties (e.g., oscillation) in order to explain these physical phenomena. As commentators like Adickes have noted, however, Kant's speculations in empirical physics are highly suspect.[72] Given these problems and Kant's later reconceptualization of the ether in the *OP,* I am equally skeptical of interpretations of Kant's transition project and the gap problem that require the ether to play an explanatory role within empirical physics (*viz.* the first strand of interpretation discussed in §0.4). When it comes to body formation, we can accept Kant's claim *that* physical bodies (or substances) emerge from the ether (or Substance), without having to accept his account of *how* they do so.[73] Much as Kant says in the General Remark on the Dynamics, explaining how physical bodies are formed goes beyond what is possible *a priori* since it requires an *a posteriori* investigation of nature. Ernst Moses Marcus, an early commentator on *OP*, makes a useful distinction between Kant's 'ether thesis' and his 'ether hypothesis.'[74] The ether thesis holds that the ether cannot be immediately perceived and that its existence as well as its properties can only be determined *a priori*. The ether hypothesis, in contrast, concerns the ether's purported role in determinate physical processes. One can accept the ether thesis, however, while rejecting the ether hypothesis. On this point I agree. *OP* is most valuable not for its scientific speculations, but rather for its philosophical arguments. In the *Übergang* section, Kant himself downplays the role the ether might play in explaining specific physical phenomena, holding instead that its existence is demonstrable as a condition of experience. As he says, 'Caloric is *actual*: it is not a material feigned for the sake of the explanation of certain phenomena, but rather, a material demonstrable from a universal principle of experience' (*OP* 22:551).[75]

In *OP*, Kant has developed a new concept of matter, which, though consistent with his conception of Substance in *CPR*, is far richer in detail since it makes fully explicit the radical idea of identifying the notion of a dynamic totality of forces with the notion of a persistent substrate of material qualities and changes. Although Kant calls the ether by many names, as the quote that starts this chapter makes clear, this material totality's name matters not, but rather only its causal-nomological functions. As I will argue in the next chapter, these various functions are not independent of one another (as some commentators hold) but are rather mutually supporting.

Assuming that Kant's post-Critical conception of the ether is simply a radicalized and more sophisticated version of his Critical conception of Substance, one can notice a significant expansion of the latter's functions. Substance is no longer merely a sempiternal and omnipresent substrate, but is also a substrate that is constituted by attractive and repulsive forces. This substrate is necessary for the sensible properties of bodies while at the same time possessing none of these sensible properties. As I will argue in the next two chapters, although the concept of this substrate can be characterized (both positively and negatively) through the categories and is ultimately applied for experience through the categories in their role as Principles of Pure Understanding, it is not itself a category.

At the same time, one should not take this to suggest that the ether is only *conceptual*. As I will argue in the next chapter, Kant believes the ether is *actual* (i.e., the concept of the ether has an extension). Using Kant's own terms from the Critical period, the ether is *objectively real*. In the Principles, Kant holds that the categories are objectively real just in case something answers to these concepts in outer sense.[76] I believe that the concept of the ether is objectively real insofar as the ether exists as an omnipresent plenum of force in outer sense. Much of the disagreement in the secondary literature (covered in §0.4) centers around the question of whether the ether is a concept (formal) *or* actual (material). Although most commentators take this as an exclusive disjunction, I believe both disjuncts are true. As I will argue over the next two chapters, the ether has both a formal as well as a material role to play in Kant's post-Critical theory of experience. My inclusive interpretation also allows one to account for passages that seem contradictory under the exclusive view. Consequently, my view should be preferred according to my first principle of interpretation (from §0.1).

Returning to my diagnosis of the gap problem from the previous chapter, assuming that Kant's argument for Substance in the Analogies fails, and given the close connections between Kant's Critical conception of Substance and his post-Critical conception of the ether, it is important to examine Kant's argument for the latter. Whereas this chapter has focused on the development of Kant's conception of the ether prior to the *Übergang* section, we must now also examine what use Kant makes of his new conception of the ether in the Ether Deduction itself.

NOTES

1. For a similar description that also captures some of the other properties mentioned below, see *OP* 22:550.
2. Wolfgang Bayerer, "Ein Vershollenes Loses Blatt Aus Kants Opus Postumum," *Kant-Studien* 58, no. 3 (1967): 284 (line 27). Likewise, in the *Übergang* section, Kant describes the ether as 'universally distributed' and 'sempiternal' [*alldaurend*] (*OP* 21:584).
3. See *OP* 22:194 and 22:378.
4. See *OP* 22:239–242. Lord holds that in *OP* attraction and repulsion are secondary forces caused by the (unexplained) primary forces of the ether. See Lord, *Kant and Spinozism,* 158. I see no textual reason to make this distinction (and Lord provides none), though there is sufficient textual evidence to suggest that mechanical forces are secondary forces caused by the dynamic forces (original attraction and repulsion) of the ether.
5. See *OP* 21:219–220 and 21:584.
6. See *CPR* A576/B604.
7. For example, see the *New Elucidation* (1755), *AK* 1:395–396. For more on Kant's pre-Critical proofs for the *omnitudo realitatis,* see Förster, *Kant's Final Synthesis,* 77–79. There are other ideas from the pre-Critical writings, relevant to our discussion, that continue to show up in the Critical and post-Critical periods. One example is the idea that physical bodies are constituted by attractive and repulsive forces. See the *Physical Mondadology* (1756), *AK* 1:476 and 1:484. In the pre-Critical period, Kant also speculates that an omnipresent and causally efficacious ether might exist as the cause of a number of different physical phenomena. See the *Only Possible Proof of the Existence of God* (1763), *AK* 2:113. In the *Physical Monadology,* he holds that ether is the matter of fire (caloric) and is underpinned by a medium of monads. See the *Physical Monadology, AK* 1:486–487. Kant identifies light-ether and caloric in *A Brief Account on Some Reflections on Fire* (1755), *AK* 1:377. For a discussion of some of these passages, see Friedman, *Kant's Construction of Nature,* 134–137. Carrier goes so far as to argue that Kant subscribed to dynamism and plenism throughout his career. See Carrier, "Kraft und Wirklichkeit," 209.

8. See *CPR* A580–583/B608–611 as well as Kant's subsequent criticisms of the standard proofs for the existence of God.
9. Although my focus will be on Kant's published work from the Critical period, one can also find mentions of the ether in his unpublished work from this same period. For example, see Reflection 45a (*AK* 14:406) and 50 (*AK* 14:443).
10. See *CJ* 5:224. Förster notes that Kant seems to change his mind on this issue between the second and third editions of *CJ*. See Förster, *Kant's Final Synthesis,* 29–30.
11. *CJ* 5:348–349.
12. As mentioned in §0.4, I believe Friedman errs by insisting on such a reduction.
13. Although Kant's most extensive discussion of the ether is in the General Remark on the Dynamics, he does mention it elsewhere in *MFNS*. In the main part of the Dynamics, Kant seems to endorse Newton's claim that the ether could gradually be transformed from a material with no weight to a material with weight (*MFNS* 4:515). He mentions a 'cosmic matter' [*Weltmaterie*] through which original attraction might operate (4:518). He also makes oblique references to light-ether (4:520) as well as caloric or heat-ether (4:522). As we will see below, he also mentions it in the Phenomenology. Although Friedman believes that Kant is committed to the existence of the ether in *MFNS,* in the passages cited, Kant seems to think of it as nothing more than a plausible (though preferable to atomism) explanation for the phenomena under discussion. See Friedman, *Kant's Construction of Nature,* 197 and 210. As we will see in the next section, Kant's post-Critical conception of the ether is not of a material medium and it cannot be reduced to Modern conceptions of light-ether or heat-ether.
14. See the Introduction (§0.3) for the circle that this view generates.
15. For further discussion of the relationship between the ether and cohesion, see the *Danziger Physics* (*AK* 29:146) and *Correspondence* 11:362 and 365.
16. As we will see in the next chapter (§3.4), this distinction plays an important role in explaining why the ether is only transcendentally necessary (for possible experience) though not metaphysically necessary (in itself).
17. See *MFNS* 4:502 and 532–533.
18. As mentioned in §0.3, for Friedman, another distinct advantage for relative impenetrability (dynamical) over absolute impenetrability (mechanical) is that the former allows for well-defined velocity at the turn-around point. See Friedman, *Kant's Construction of Nature,* 116.
19. See *CPR* A434–441/B462–469 and A524–527/B552–555.
20. See *OP* 22:269.
21. See *OP* 22:194.
22. For example, see *OP* 22:212, 21:218, and 22:474.
23. See *OP* 22:481.
24. For an explanation of the corpuscularian hypothesis see Robert Boyle, "The Origin of Forms and Qualities According to the Corpuscularian Philosophy," in *The Selected Philosophical Papers of Robert Boyle,* ed. M.A. Stewart (Indianapolis: Hackett, 1991), 1–96.
25. See Westphal, *Kant's Transcendental Proof of Realism,* 192–194.
26. This is true of Kant's theory even if his argument for this theory rests upon a non-sequitur (see §0.3). Even so, one might object that the fundamental forces are equally "occult" since 'they can in no way be conceived' (*MFNS* 4:513). Pollok claims, however, that this only means that the concepts of the fundamental forces cannot be derived from any other concepts. See Pollok, *Kants Metaphysische Anfangsgründe,* 297.
27. See *OP* 21:310. See also *OP* 21:274 where Kant ties the forces of the ether back to original attraction and repulsion. Unlike Emundts (see §0.4), I do

believe Kant makes an effort to tie his conception of the ether back to his dynamic conception of matter in *MFNS*. One might also note, however, that the ether, in *OP*, is not locomotive. Being locomotive is an essential property of matter in *MFNS*.

28. See also *OP* 21:227 where Kant claims that the primordial dynamic motion of the ether is necessary for all mechanical motion in space. It is interesting to note that Kant's theory does not require any external cause (e.g., a Cartesian God) to preserve the quantity of motion in the world.
29. See *OP* 21:378–379. Although Förster thinks that Kant can avoid the circle that faces his theory of density by distinguishing the contact force of cohesion from the penetrating force of original attraction (or gravitation), insofar as the former is only apparent and depends partially on the latter, it is unclear how the circle is to be avoided. See Förster, *Kant's Final Synthesis,* 41–45.
30. Consequently, I cannot agree with Emundts that Kant's ether theory offers a solution to the circularity problem. Even so, I do not think this constitutes a gap in the Critical philosophy for reasons I outlined in §0.4. For Emundts' account of the solution, see *Kants Übergangskonzeption,* 103–106.
31. Westphal claims that ultimately Kant needs three kinds of attractive force: (1) original attraction, (2) universal gravitation, and (3) cohesion. See Westphal, *Kant's Transcendental Proof of Realism,* 197.
32. See *MFNS* 4:518, 541 and *Correspondence* 11:376–377.
33. See *OP* 21:378.
34. Pollok notes that Kant himself is not always consistent in the way that he characterizes the relationship between fundamental forces and matter in *MFNS*. In Proposition 5 of the Dynamics, for example, he seems to fluctuate between fundamental forces constituting matter and being accidents of matter (*MFNS* 4:508–510). See Pollok, *Kants Metaphysische Anfangsgründe,* 281–282.
35. See *MFNS* 4:498–499.
36. See Westphal, *Kant's Transcendental Proof of Realism,* 201–204.
37. See Pollok, *Kants Metaphysische Anfangsgründe,* 347–348.
38. See *OP* 21:523.
39. See *OP* 22:271.
40. Translation is mine.
41. See also the discussion of body formation in James McCall, "A Response to Burkhard Tuschling's Critique of Kant's Physics," *Kant-Studien* 79, no. 1 (1988), 57 and "Metaphysical Foundations and Ponderomotive Nature," *Kant-Studien* 96, no. 3 (2005): 308. Although the former account is fairly close to the one I present above, the latter account relies heavily on special relativity and quantum mechanics which makes it somewhat anachronistic.
42. Jeffrey Edwards and Martin Schönfeld, "Kant's Material Dynamics and the Field View of Physical Reality," *Journal of Chinese Philosophy* 33, no. 1 (2006): 109. Paolo Pecere also recognizes the connection between the ether as a dynamic continuum and physical fields. See Paolo Pecere, *La filosofia della natura in Kant* (Bari: Pagina, 2009), 738–739. For a discussion of this book (in German), see Basile, *Kants Opus postumum,* 296–299. McCall updates Kant's view for contemporary science by discussing the conversion of kinetic energy into rest energy or rest mass. Using the formula $E = MC^2$ from the special theory of relativity, one can see that the mass of a body is proportional to its kinetic energy. McCall says that 'what for Kant was an *Anhaefung von Waermestoff* [accumulation of caloric] is for us a continuous conversion of kinetic energy over space and time into rest-energy.' As mentioned above, although McCall's view is somewhat anachronistic, this analogy does strike me as being in the spirit of Kant's project. See McCall, "Metaphysical Foundations and Ponderomotive Nature," 308.

43. Edwards and Schönfeld, "Kant's Material Dynamics and the Field View of Physical Reality," 116.
44. Hanna, *Kant, Science, and Human Nature,* 149. Kant's influence on the development of contemporary field theory is well documented. For example, see Mary Hesse, *Forces and Fields* (New York: Philosophical Library, 1962), 170–180, William Berkson, *Fields of Force: The Development of a World View from Faraday to Einstein* (London: Routledge, 1974), 22–34, and Friedman, *Kant's Construction of Nature,* 97. Although Boscovich's theory of material *points* imbued with forces directly influenced Faraday, the actual ontology that Faraday adopts is closer to Kant's own. Faraday takes force as ontologically fundamental and so views fields of force as *continuous.* For more on the relationship between Boscovich and Faraday, see Hesse, *Forces and Fields,* 201 and Berkson, *Fields of Force,* 50.
45. See *OP* 22:239–242.
46. Although in the Dynamics of *MFNS,* the attractive and repulsive forces are not internally moving (i.e., oscillating), they are still the foundation of matter. For other references to Kant's dynamic theory of matter over the course of Kant's transition project, see *OP* 21:373, 21:223–224, and 22:437. Howard Caygill adopts a view toward matter in *OP* similar to my own. See Howard Caygill, "The Force of Kant's Opus Postumum: Kepler and Newton in the XIth Fascicle," *Angelaki* 10, no. 1 (2005): 37 and 40.
47. See *MFNS* 4:524.
48. Bayerer, "Bemerkungen zu einem neurerdings näher bekannt gewordenen Losen Blatt aus Kants Opus Postumum," *Kant-Studien* 72, no. 2 (1981): 131 (lines 39–40).
49. See also *OP* 22: 594.
50. See *OP* 22:268–269.
51. Violetta Waibel argues that Kant's characterization of the *dynamical* ether is initially negative precisely because it lacks any of the properties that *mechanical* matter possesses. See Violetta Waibel, "Des principes régulateurs qui sont en même temps constitutifs," in *Années 1796–1803. Kant. Opus postumum. Philosophie, Science, Ethique et Théologie. Actes du 4e Congrès international de la Société d'études kantiennes de langue française. Lausanne: 21–23 octobre 1999. Sous la direction de Ingeborg Schüßler,* ed. Christophe Erismann (Paris: Vrin, 2001), 155–156. For discussion of this article (in German), see Basile, *Kants Opus postumum,* 293.
52. Adickes also contrasts the positive properties of material objects with the negative properties of ether. See Adickes, *Kants Opus postumum,* 326.
53. Förster describes the *Elementarsystem* in a similar way. See Förster, *Kant's Final Synthesis,* 113–114.
54. See *OP* 21:584. Here I am following Förster, *Kant's Final Synthesis,* 97.
55. Translation is mine.
56. Translation is mine.
57. See *OP* 22:559.
58. For an excellent discussion of the various problems Kant faced in trying to formulate the *Elementarsystem,* see Förster, *Kant's Final Synthesis,* 11–23.
59. The distinction between Kant's negative and positive characterization of the ether is often overlooked in the literature. Beiser, for example, suggests that the fundamental forces that constitute the ether can themselves be classified according to the categorical properties that substances possess. See Beiser, *German Idealism,* 186. I believe that this is true only of the effects of these forces, but not the forces themselves.
60. As mentioned in §0.4, Förster identifies the ether with a transcendental ideal. It functions as a regulative principle for the unification of experience. See

Förster, *Kant's Final Synthesis,* 91–92. Lehmann adopts a similar view, but for a different reason. He believes that this is the most that can be hoped from the Ether Deduction given weaknesses within the proof itself. See Lehmann, *Interpretation der Philosophie Kants,* 193 and 255–256.

61. Emundts argues that Kant's proof for the existence of the ether actually begins in the *Elementarsystem* under the category of modality. See Emundts, *Kants Übergangskonzeption,* 158. For example, Kant suggests that the moving forces of matter fall under the category of necessity. Given the omnipresence and sempiternality of the ether, Kant describes it as 'necessity in appearance *(Perpetuitas est necessitas phaenomenon)*' (*OP* 22:188).
62. See *CJ* 5:370–371.
63. Kant uses the watch example. See *CJ* 5:374.
64. See *CJ* 5:374.
65. See First Introduction to *CJ* 20:235.
66. See *CJ* 5:360. Hannah Ginsborg discusses this issue at some length and offers a *normative* account of explaining organic nature within the context of *CJ*. See Hannah Ginsborg, "Kant on Understanding Organisms as Natural Purposes," in *Kant and the Sciences,* 231–258.
67. In this respect, I agree with Mathieu that Kant aims to establish constitutively the systematic unity of nature though I do not think the inadequacy of Kant's conception of purposiveness in *CJ* constitutes the gap in Kant's Critical philosophy for the reasons given in the Introduction (§0.4).
68. Guyer, "Organisms and the Unity of Science," in *Kant and the Sciences,* 260.
69. See *OP* 21:211–213.
70. Using *CJ,* Lord argues that if the ether were able to account for organic nature it would be equivalent to Spinozistic substance (*CJ* 5:393). See Lord, *Kant and Spinozism,* 170–173. I have already offered independent reasons, however, for thinking that Kant's conception of the ether is not identical to Spinoza's conception of substance (see §1.1). Kant's more general point in *CJ,* however, is that matter cannot possess intentionality and so cannot produce organisms (natural purposes). Admittedly, there are several passages throughout *OP* that suggest that this remains Kant's view (e.g., see *OP* 22:210, 22:506–507, 22:548) and these passages inform some of the interpretations in the secondary literature (e.g., see Lord, *Kant and Spinozism,* 172 and Beiser, *German Idealism,* 189). These interpretations, however, cannot overcome the problem that Guyer raises. According to my third principle of interpretation, my view should be preferred insofar as it can actually solve the philosophical problem that Kant faces. Although I agree that the ether cannot possess intentionality, in the *Elementarsystem,* Kant suggests that organic nature can still possess final causes. We can only *understand* (epistemic) these final causes, however, through an analogy with our own intentionality. If inorganic and organic nature can be understood as emerging from the ether, this will allow Kant to unite both inorganic and organic nature into one world system. The latter is vital if Kant is to establish the systematic unity of nature, a goal Kant has in both the Critical and post-Critical periods.
71. Lord holds that the human being as (1) an organic body and (2) a free subject cannot be produced by the ether. See Lord, *Kant and Spinozism,* 174. Although I challenge (1), I certainly agree with (2). In addition, I will argue in §5.5 that the emergence of both organic and inorganic bodies requires the collusion of the ether's physical activity with the subject's mental activity.
72. For an overview of the different opinions on the value of Kant's scientific claims in *OP,* see Basile, *Kants Opus postumum,* §1.2.6, 1.4, 1.5, and 2.1.2.
73. Although I accept an emergence view of the relationship between substances and Substance, at this point a comparison with the contemporary discussion

in supervenience might be helpful. One can hold the *supervenience* thesis *that* there is no difference in A-properties without a difference in B-properties without holding the *superdupervenience* thesis that explain *how* the relationship works or *why* there is no difference in A-properties without a difference in B-properties. The term "superdupervenience" was coined by William Lycan, "Moral Facts and Moral Knowledge," *Southern Journal of Philosophy* 24, (1986), 92.

74. Ernst Moses Marcus, "Kant und der Aether," *Frankfurter Zeitung und Handelsblatt. Literaturblatt* 18–19, (1921).
75. Kant will say almost exactly the same thing in the version of the Ether Deduction I reconstruct from *Übergang* 11 in the next chapter.
76. See *CPR* B291–293. Although Friedman argues that this conflates the conditions for the objective reality of pure concepts with those of empirical concepts, the textual evidence in favor of my view (which I share with Förster) is fairly compelling. For the alternative view, see Friedman, *Kant's Construction of Nature,* 573–578. Even if the conditions for the objective *reality* of pure concepts and empirical concepts are the same, pure concepts are objectively *valid* (i.e., conditions for the possibility of objects) in a way that empirical concepts are not. See *CPR* A158/B197.

3 Kant's Ether Deduction

> There is only one space, one time, and one matter, in which all motion is to be found. The real and objective principle of experience which, in its form, amounts to a unified whole, leaves no space (inside or outside itself) unfilled. It contains all moving forces. This composite is not locomotive; nor is it a body. The beginning of its motion is its own eternity. . . . Caloric is the basis for the unified whole of all moving forces of matter (the hypostatized space itself, as it were, in which everything moves); the principle of the possibility of the unity of the whole of possible experience.
>
> —Kant, from *Übergang* 2 (*OP* 21:224)

3.0 INTRODUCTION

This chapter is an attempt to reconstruct a valid and arguably sound version of Kant's Ether Deduction from the *Übergang* [transition] section of *OP*. In accordance with the above quote, I will argue that Kant's Ether Deduction establishes the *actuality* of the ether as a *material* transcendental condition for the unity of experience. The chapter is divided into four sections. The first section discusses the structure of the *Übergang* section and its function within Kant's transition project. The second section offers a translation of an extended selection of passages from *Übergang* 11 as well as a formal reconstruction of Kant's argument there. Section three examines the main steps of this reconstruction of the Ether Deduction in light of the Critical project so that one might see that Kant's Ether Deduction is a significant post-Critical development of the Critical-era philosophy. The final section evaluates the success of Kant's Ether Deduction first with regard to my particular reconstruction but then also in terms of what Kant might have *intended* with the Ether Deduction compared to what he actually *achieved*.

3.1 THE STRUCTURE AND FUNCTION OF THE *ÜBERGANG* SECTION

Versions of the Ether Deduction are scattered throughout a series of fourteen *Übergang* drafts. Fortunately, Kant numbered these drafts in his manuscript and we can be fairly certain that they were all written between May and August 1799. Although not all of the *Übergang* drafts contain a version of the Ether Deduction, the ones that do bear striking similarities to one another. This is particularly true of their shared proof strategies and major premises. At the same time, they each contain their own unique differences in style and structure.

There are amanuensis copies of *Übergang* 9–11. As mentioned in the Introduction (§0.1), amanuensis copies were one of the last steps before publication, which means that Kant, at least at some point, thought that these three drafts were the closest of all to being his considered position. The Cambridge edition of *OP* translates the amanuensis copies, but these copies are missing sections of text from the original. I have chosen to translate an extended section from the original version of *Übergang* 11 since I think it is the most compelling formulation of the Ether Deduction. The first two paragraphs of the section I translate (the proof itself) are nearly identical to a section in the amanuensis copy that is translated in the Cambridge edition.[1] I have also included a couple of notes from the original that I think are helpful in understanding the proof.

A recently published letter written by Friedrich Theodor Rink, on Kant's behalf, to Jeronimo de Bosch reinforces the idea that Kant thought his project was nearing completion.[2] The letter was likely written on July 20th, 1799.[3] In it, Rink explains why Kant has tasked him with writing the letter. According to Rink, Kant does not have time to write Bosch since he is too busy compiling his new work [*in concinnando nouo libro opera dirent*] which implies a fairly late stage in the writing process.[4] Since Kant approved this letter before it was sent (through the addition of his inscription), we can be fairly certain that the letter accurately represents Kant's view on the state of the manuscript.[5]

There are good textual reasons (e.g., correspondence and amanuensis copies) for giving the *Übergang* section pride of place in an interpretation of the gap and Kant's transition project. In fact, Kant entitles one *Übergang* draft: 'The existence of the ether [*Wärmestoffs*] as the **highest principle** of the transition from the metaphysical foundations of natural science to physics' (*OP* 21:600).[6] Assuming that the ether is the "highest principle" of the transition, and that the redescribed transition fills the gap in the Critical philosophy (as the letters to Garve and Kiesewetter suggest), one should try to find a gap in the Critical philosophy that a successful Ether Deduction might fill. I will continue to argue that the parallels between the Analogies and the Ether Deduction, the Substance and ether, are too strong to ignore. This provides good philosophical reason, I believe, for locating the gap in the Analogies of Experience.

Förster attempts to reconstruct the Ether Deduction, but instead of focusing on a specific formulation of the deduction in one of the fourteen *Übergang* drafts, he tries to synthesize all of the formulations as well as material from the Critical era into one comprehensive proof. Although Förster's goal is admirable from a systematic perspective, the result of his reconstruction is a cumbersome affair with eleven premises and thirteen steps divided into a subjective and an objective phase.[7] He admits that most of the material for the objective phase is only implicit in the *Übergang* section, which leads him to cull the premises for this phase from the Critical-era works instead of from the *Übergang* section itself. Without evaluating the soundness of Förster's reconstruction, it is important to note that Kant's *own* formal proof strategy in the *Übergang* drafts is dropped in Förster's version. I believe this is a great loss and in my reconstruction of Kant's proof I will strive to retain its formal structure.

It is not very surprising that Förster faces problems synthesizing the different formulations of the Ether Deduction given the fact that Kant sometimes argues in very different ways for the same conclusion. Guyer lists at least four different types of arguments for the ether in the *Übergang* section. At different points, Kant argues (1) that the ether is necessary for the unification of space and time, (2) that the ether is required to preclude or rule out the experience of empty space, (3) that the ether is the medium for transmitting the force necessary for perception, and (4) that the ether is the original cause of motion.[8] Of these four types of argument, my reconstruction will focus on (2), although, as I will argue below, I believe (2) is also closely related to (1). As we will also see, Kant's argument touches on (3) and (4) in the process of establishing (2). Much as in the Third Analogy, Kant argues that the ether is necessary to rule out the experience of empty space.[9] In this context, the specific problem with empty space that needs to be ruled out is that experience would be "broken off." Consequently, the unity of experience in space and time requires that Kant rule out the experience of empty space.

As Kant says repeatedly throughout the *Übergang* section, the Ether Deduction is meant to effect a transition [*Übergang*] from the metaphysical foundations of natural science to physics.[10] In this respect, it can be seen as a major post-Critical development of his Critical project. Whereas in *CPR* Kant attempts only to elucidate the *a priori* formal transcendental conditions for experience (*viz.* space, time, categories, apperception), and *MFNS* attempts only an application of Critical principles to the objects of physics by categorically determining the empirical concept of matter, Kant is here attempting to prove *a priori* the *existence* of a material which is itself a *transcendental* condition for experience as well as the ultimate object of physics.

Although Kant's argument for "matter everywhere" or what I have called "Substance" in the Third Analogy (see §1.1) anticipates his argument for a transcendental material condition of experience in the Ether Deduction, with only the category of substance, Kant lacks the conceptual resources necessary to account for *both* Substance (sempiternal and omnipresent) as well as substances (relatively persistent and spatially discrete) in the Analogies. This creates a dilemma for his theory of substance in *CPR* which consequently exposes a gap in the Critical philosophy, specifically in the transcendental part (i.e., the Transcendental Analytic) of the metaphysics of nature. Furthermore, given the likely application of the category of substance only to what I have called "substances," it would seem that Kant's argument for Substance in the Analogies must be considered a failure. By building on his argument from the Third Analogy, Kant's post-Critical Ether Deduction delivers not only the existence of a transcendental material condition but also an *a priori* concept of this material totality, importantly different from the category of substance, and necessary to avoid the dilemma he faces in the Analogies. Although he will not explain how the gap is bridged until *Convoluts* 10–11 (see Chapter Four), it is here in the Ether Deduction that Kant takes the first steps toward bridging the gap.

Given the problems I have pointed out with Förster's reconstruction of the Ether Deduction, I will not try to consider all fourteen drafts of the *Übergang* section. Instead, I will focus on what seems to me to be Kant's most compelling formulation of the deduction in *Übergang* 11. As in many of Kant's proofs, the individual steps are not immediately clear and the premises are not always given sequentially. For the sake of clarity, in the next section I will quote Kant's formulation at length and then attempt to reconstruct the proof formally, while at the same time remaining as faithful as possible to Kant's original text.

3.2 *ÜBERGANG* 11 AND ITS FORMAL RECONSTRUCTION

From *Übergang* 11:

> If it can be proven that the unity of the *whole* of possible experience is founded upon the existence of such a material [ether] (with its stated properties) so its actuality is also proven, not **by** experience but rather *a priori* **for** experience, merely from the conditions of its bare possibility. For the moving forces of matter can only harmonize [*zusammenstimmen*] into a collective/universal unity of perception in one possible experience insofar as the subject [is] affected by them externally, and united in one concept, internally affects itself.
>
> (*OP* 21:572.16–24)
>
> Now the concept of the whole of outer experience presupposes all possible moving forces of matter combined in collective unity, specifically in *full* space (because empty space either inside or outside a physical body is no object of possible experience). It [concept of the whole of outer experience] presupposes, however, also a continuous motion of all matter which effects the *subject* as a sensible thing, because without this

motion, i.e. without the agitation of the sense organs, as this motion's effect, there would be no perception of sensible objects and with that no experience occurs, which is nothing but the form appropriate to that [perception]. —Therefore, there is a material, continuously and unlimitedly expanded in space, which is itself the particular agitating material as an experiential thing (although without empirical consciousness of its principal), that is to say the caloric is *real* and not merely a feigned material for the explanation of certain phenomena, rather, an *a priori* given material, analytically provable, by the principle of identity, from a universal principle of experience.

(*OP* 21:572.25–27/573.1–13)

Note Two: The proof of the above proposition is unique in its type which must not amaze since the object of *one* collective experience to which the concepts of the moving forces of matter refer, is itself unique. Basically, we proceed indirectly, i.e. to show the truth of the proposition through the impossibility of its opposite: namely because in the opposite case empty space would be a possible object of experience which is itself contradictory.

(*OP* 21:574.14–20)

[Right-Hand Marginal Note]: Deduction of an empirical concept from the subjective principle of the possibility of *one* experience which constitutes an *a priori* cognition according to the principle of identity.

(*OP* 21:573.15–17)[11]

Kant states at the outset that the unity of experience is *founded* upon the ether, or as he says more clearly at the beginning of the second paragraph it *presupposes* the ether. The latter formulation is logically useful, because it tells us that the ether is meant to be a *necessary* condition for the unity of experience. The structure of the argument, then, is P $\rightarrow$ Q, where we are to assume that there is a unity of experience (P) and derive the existence of the ether (Q). Although something very much like (Q) is *anticipated* by Kant's arguments in the Analogies (i.e., Substance), it cannot be *established* by Kant's arguments in the Analogies given the dearth of conceptual resources available to him (i.e., no *a priori* concept of Substance). Consequently, in *CPR,* Kant seems limited to only four necessary and jointly sufficient conditions for the unity of experience: space, time, categories, and apperception. I represent these collectively as (R).

Although (R) represents *necessary* conditions for the unity of experience, how can it represent *sufficient* conditions? Surely (R) cannot justify the *details* of experience since the manifold of empirical intuition is *a posteriori* and contingent.[12] I do not want to claim that the transcendental conditions of experience are sufficient to establish the *details* of experience, though they are sufficient to establish the *general* conditions under which experience is unified. Although Kant later takes the unity of experience as given (a point I will return to below), at various places in the Transcendental Analytic (e.g. the Analogies), his arguments aim to establish this unity by providing the necessary and sufficient conditions for it.

Assuming that these four necessary conditions are supposed to be jointly sufficient for the *unity* of experience, one should represent the relation between (P) and (R) as a biconditional: $P \leftrightarrow R$. Kant argues in *Übergang* 11 that the Critical set of conditions (R) is, in fact, not sufficient for (P). It is only through the conjunction of (Q) and (R) that one might establish the necessary and sufficient conditions for (P): $P \leftrightarrow (Q \wedge R)$. A material transcendental condition must be established to insure the unity of experience. If Kant's argument for this material condition in the Analogies fails, then one must look to a post-Critical solution to this Critical-era problem. The proof of $P \rightarrow Q$, if successful, seems then to serve two purposes. (1) It exposes a gap in the Critical project since (R) is no longer sufficient for (P). (2) It potentially fills this gap if it can show both that (Q) is a necessary condition for (P) and jointly sufficient with (R). I will argue that (1) reflects the failure of Kant's arguments in the Analogies and that although (2) succeeds in its claim of necessity it fails to meet the standards of sufficiency within the framework of the Ether Deduction. After formally reconstructing Kant's proof, I will proceed in the next section to motivate each of its steps in light of the Critical project. Kant's proof of $P \rightarrow Q$ can be reconstructed as follows:

1. (P) There is a unity of experience (21:572.16–24).
2. The subject comes to experience external objects only insofar as these objects affect the subject by means of the moving forces of matter, and as the consequent perceptions are united by an act of apperception into one concept (21:572.16–24 and 21:572.25–27/573.1–13).
3. Empty space cannot affect the subject (21:572.25–27/573.1–13).
4. There must be an actual object that is an analytic consequence of the concept of the unity of experience (21:572.25–27/21:573.1–13 and 573.15–17).
5. From (4), The actual object that is an analytic consequence of the concept of the unity of experience can be either (Q) the ether with its stated properties, or (synthetic ¬Q) a mechanical whole of physical bodies containing or separated by empty space [Theorem Introduction] (21:572.25–27/573.1–13 and 21:574.14–20).

    i. [Assume (5) synthetic ¬Q for *reductio ad absurdum*] A mechanical whole of physical bodies containing or separated by empty space is the actual object that is an analytic consequence of the concept of the unity of experience.
    ii. From (1) and (2), everything external, comprehended within the unity of experience, must be capable of affecting the subject.

iii. From (i) and (ii), empty space can affect the subject.
iv. Contradiction: (3) and (iii).

6. (Q) From (i)–(iv), the ether with its stated properties is the actual object that is the analytic consequence of the concept of the unity of experience [Indirect Proof] (21:574.14–20).

3.3 EXAMINING THE ETHER DEDUCTION IN LIGHT OF THE CRITICAL PROJECT

Having reconstructed Kant's proof formally, in this section I will explain each of the steps using material *primarily* from *Übergang* 11, but when necessary, for the sake of clarity, I will incorporate material from Kant's other *Übergang* drafts. Most importantly, however, I will examine each of Kant's main steps in light of the Critical project in the hopes that one might see that the Ether Deduction constitutes an important post-Critical development of the Critical-era philosophy.

One might find the starting point of Kant's argument a bit surprising. From where does the unity of experience come? Kant's answer can be found in the Transcendental Ideal section of *CPR* discussed in Chapter One (§1.3). Quite simply, it is *given*. As Kant says:

> But because that which constitutes the thing itself (in appearance), namely the real, has to be given, without which it could not be thought at all, but that in which the real in all appearances is given is the one all-encompassing experience.
>
> (*CPR* A581–582/B609–610)

Kant's beginning premise of *Übergang* 11 is consistent with his views from the Critical period and also reflects the unity of time-determination that Kant tries (and fails) to transcendentally justify in the Analogies. The one experience in which all perceptions are represented as in thoroughgoing and law-like connection (Transcendental Deduction), the unity of apperception with regard to all possible empirical consciousness (Analogies), the one all-encompassing experience (Transcendental Ideal), or the unity of the whole of possible experience (*OP*), is given as that through which appearances are real. I believe that Kant is referring to the same thing in each of these formulations, namely, the unity of experience.[13]

As mentioned above, the Critical philosophy contains a gap that the Ether Deduction helps to expose. Even if there is one all-encompassing experience or unity of experience, in what way can the *real* of all appearances be conveyed through this unity of experience? If I am right about the Analogies, Kant's argument for the totality of empirical reality or Substance fails within the context of the Critical project. Kant does, however, attempt to redress this failure in *Übergang* 11 of *OP*. This is to say that the Ether Deduction is itself an argument for the totality of empirical reality and consequently exposes this gap within the Critical philosophy. Although the totality of empirical reality can only be presupposed as the material condition for the possibility of any object of experience in the Transcendental Ideal, the Ether Deduction provides the post-Critical argument for this Critical-era presupposition.[14]

Kant also makes the connection between the Transcendental Ideal and the Ether Deduction explicit in the *Übergang* section. At one point, he claims that the principle of thoroughgoing determination contains a proof for the ether and offers the definition '*existentia est omnimoda determinatio*' [existence is thoroughgoing determination]' (*OP* 21:577.20–21).[15] Much as in the Transcendental Ideal, applying the principle of thoroughgoing determination to objects presupposes the sum total of all predicates, so that the predicates of the object might be determined relative to such a sum total. In *OP*, Kant considers the ether to be this sum total. At first blush, this might seem somewhat strange, but here it is important to remember the *negative* characterization of the ether discussed above (§2.2). Kant sees the ether as something which makes all of the specific determinations of bodies possible (e.g., ponderable, cohesible, etc.), without itself possessing any of these properties. The ether itself can only be negatively characterized relative to the properties that physical bodies or substances possess, but can be positively characterized, through its effects in the physical world, relative to these same properties. In the same way, the ether is here assumed for the application of the principle of thoroughgoing determination to the objects of experience, though it is not itself subject to this principle.[16] Unlike *CPR*, where Kant simply *presupposes* the sum total of possible predicates, however, in *OP* Kant thinks that he can offer a *proof* for this sum total *a priori*. In any case, Kant's opening premise is consistent with both his Critical and post-Critical views on experience:

1. (P) There is a unity of experience.

In 21:572.25–27/573.1–13, Kant states that the moving forces of matter collectively constitute a condition for all external experience since they affect the subject, making perception and, consequently, outer experience possible. In the Preface to *MFNS*, Kant claims that the fundamental characteristic of an object of outer sense is motion since only by means of motion can the subject be affected.[17] As I argued above (§1.2), in this context Kant seems to mean locomotion. If the ether is omnipresent, however, how can it be locomotive? In the quote that opens this chapter, Kant explicitly states that the ether is not locomotive. Would this not preclude the ether from affecting the subject? Although a full answer to this question will have to wait until later chapters (§4.1 and §5.5), let it suffice to say that Kant clearly makes room for both *dynamical* as well as *mechanical* affection in *OP*. One might also note that although Kant has not yet adopted a dynamical ether theory in *MFNS,* he does believe that mechanical forces depend on dynamical forces.[18] Kant develops this view in *OP* (see §2.2). The locomotion of bodies (mechanical) depends upon the internally moving forces of the ether (dynamical). It is ultimately by means of these internally moving forces that locomotive physical bodies affect receptive subjects in sensibility.[19] Even though aspects of Kant's considered view on affection can be found in *CPR* and *MFNS,* the idea that there is a dynamical basis for affection is only fully developed in *OP.*

Whereas Kant states at 21:572.16–24 that the moving forces of matter are an *external* condition for experience, he mentions in the same breath that there is an *internal* condition for experience. The subject is affected from without, but unites the consequent perceptions in one concept through self-affection. This spontaneous act of self-affection is *apperception,* a formal cognitive condition of experience, found in *CPR* and made use of here again in *Übergang* 11.[20] Although I will have much more to say about Kant's post-Critical view on affection in the final chapter, what has been said is sufficient to introduce the Ether Deduction's second premise:

2. The subject comes to experience external objects only insofar as these objects affect the subject by means of the moving forces of matter, and as the consequent perceptions are united by an act of apperception into one concept.

Edwards has argued convincingly that Kant's arguments in the Analogies anticipate his argument in the Ether Deduction. [21] Specifically, the idea that empty space is not an object of possible experience appears in the Third Analogy. As mentioned above (§1.1), in the Third Analogy, Kant claims that there must be "matter everywhere" to insure the unity of the subject's spatial experience. Although Kant is using a thoroughly dynamic conception of matter in the *Übergang* 11 argument, ruling out the experience of empty space is an important aspect of both arguments and it is clear that one can find a precursor of Kant's Ether Deduction in the Analogies themselves. Much as the ether is a necessary condition for the spatiotemporal unity of experience in the Ether Deduction, so too is Substance a necessary condition for the spatiotemporal unity of experience in the Analogies.

Perhaps anticipating his own dynamical view of Substance that comes to fruition in *OP,* in a note written in his own copy of the A-edition of the First Analogy, Kant says that the alteration of substance is effected through 'moving causes' (*AK* 23:30). In the Second Analogy, Kant claims that 'where there is action, consequently activity and force, there is also substance' (*CPR* A204/B250). Finally, Kant says in the Amphiboly section of *CPR* that 'we know substance in space only through forces that are efficacious in it' (*CPR* A265/B321).

According to Guyer, there are two ways in which the ether could rule out the experience of empty space: either (1) every point in space contains a perceivable object, or (2) there is no point in space from which some object may not be perceived.[22] Of these two options, the second seems to be preferable. In ordinary experience, space certainly does not seem to be "filled up" with perceivable objects or what I have called "substances." There are often spaces in between the objects perceived in outer sense that do not themselves contain perceived objects (e.g., the space between the earth and the moon in the Third Analogy). The question is whether or not these spaces are *empty* spaces.[23] This is where the second option comes into play. The ether or "Substance" rules out the experience of *empty* space by occupying space through its moving forces even if the space that is occupied does not contain a perceivable object or "substance." Assuming that space is thoroughly occupied by moving forces, and also that these moving forces are the means by which substances affect receptive subjects in sensibility, then the omnipresence of these moving forces will guarantee that there is no point in space from which a substance might not be perceived or at least be connected to perception in accordance with natural law (via the Second Postulate).[24] Kant's claims in the *Übergang* section seem to support this interpretation:

> The transition from one object of the senses to another cannot be an experience if there is an intervening void; the two objects can be combined with each other within one experience only by means of the intermediary object of perception, which is a moving force and real material. . . . Empty but perceptible intermediary space is, thus, really a matter which in degree is imperceptible relative to our sense. . . . That by means of which space becomes an object of possible experience in general (of measure, direction, etc.) is a universally distributed, all-penetrating world-material, possessing moving forces.
>
> (*OP* 21:229)

One can certainly imagine regions of space where there may be no perceivable substances even if there are moving forces, but these substances, however remote, would still be connected with perception in accordance with natural law. The moving forces of the ether *preclude* the experience of empty space even if they do *include* the experience of any particular substance.

Kant claims repeatedly that empty space, either inside or outside a physical body, is no object of possible experience. This is not a dogmatic *metaphysical* claim (as the Rationalists were prone to make). Much as in the Third Analogy, Kant is not saying that empty space is impossible in principle. Rather, he is making the Critical *transcendental* claim that it is no object of possible experience for creatures like us. In a note attached to the above proof from *Übergang* 11 in the amanuensis copy, Kant says that 'the nonbeing of an object of perception cannot be perceived, empty space is thus not an object of possible perception' (*OP* 22:551). The truth of this assertion is absolutely central to the Ether Deduction for reasons that will become clear below. Given what is said in premise (2), however, the reasons *why* empty space is not an object of possible experience should by now be quite clear. In both *CPR* and *OP*, the subject can experience only what it can perceive. The subject can perceive only what affects it. What affects it must be an object that bears an affective relation to it. This "object" could be either a physical body which affects us by way of the moving forces of the ether or it could be the ether itself.[25] Empty space is not an object capable of exerting force (mechanical or dynamical), however, and so it cannot affect the subject. Hence, empty space is not an object of possible experience. Simply put:

3. Empty space cannot affect the subject.

What is clear from the end of 21:572.25–27/573.1–13 as well as 21:573.15–17 is that the deduction will show that something actual is analytically provable from the concept of the unity of experience. I will not dwell long on the different names that Kant calls the object of his proof, e.g., ether, caloric, or *a priori* given material. As the quote that began the last chapter makes clear, the name by which we refer to the object of this proof matters not, but rather its function which depends on it possessing certain properties. It must be a collectively moving material, continuously expanded, and constantly agitating the subject in outer sense. As I mentioned before, there also seems no point in dwelling on the different things that Kant calls the starting point of his deduction, e.g., the unity of the whole of possible experience, the universal principle of experience, or the principle for the possibility of *one* experience. What seems clear to me, however, is that Kant's starting point is the unity of experience and he means to derive his conclusion through a transcendental argument. In other words, Kant is going to show that the unity of experience would itself be impossible if the ether did not exist as an actual object. Kant's statements about analyticity and the principle of identity, however, are quite obscure. A short passage from *Übergang* 4 may help to clarify matters:

> It [the ether] is the basis of a system of moving forces which emerges analytically according to concepts, i.e. according to the principle of identity, from the principle of harmonization [*Zusammenstimmung*] with the possibility of experience in general.
>
> (*OP* 21:233.9–12)[26]

It is important to note that Kant uses the verb "*zusammenstimmen*" when he is discussing self-affection in 21:572.16–24. The role of apperception (spontaneous self-affection) in the Ether Deduction is to *harmonize* the perceptions generated by the moving forces of matter within the concept of the unity of experience. The subject begins with the unity of experience as given. Affection by the moving forces of matter results in a constant array of perceptions that the subject harmonizes within the *concept* of the unity of experience. At this point, the transcendental philosopher analyzes the concept of the unity of experience, and judges that the concept of the ether is contained within the concept of the unity of experience. In other words, the philosopher makes an analytic judgment on the basis of concepts. At a different point in the *Übergang* section, Kant emphasizes the fact that the concept of the ether is contained within the concept of the unity of experience:

> The absolute unity of possible experience is simultaneously the unity of the whole of matter, hence also the unity of the moving forces of matter that influence outer sense. Therefore, the concept of a system of agitating forces of matter lies already in the concept of the unity of experience *a priori* (before everything empirical which is just an aggregate of perceptions) as belonging necessarily to experience.
>
> (*OP* 21:595.19–20 and 596.1–4)[27]

This view is supported in 21:573.15–17, where Kant says that the deduction is a derivation of an '*a priori* cognition' of a concept *from* the unity of experience.[28] What is the intension of this concept? Kant makes oblique reference to 'its stated properties' in 21:572.16–24. I would suggest that the intension of this concept comprises the functions Kant attributes to the ether in *OP* (§2.0) as well as its (positive and negative) categorical characterization (§2.2).[29] As we will see in the next chapter, this *a priori* concept plays a vital role in bridging the gap I have identified in the Critical philosophy. Since there exists a unity of experience, and its concept *contains* the concept of the ether, the ether itself also exists. One could even say that the ether is an analytic *consequence* of the concept of the unity of experience. The below diagram may help to clarify the process:

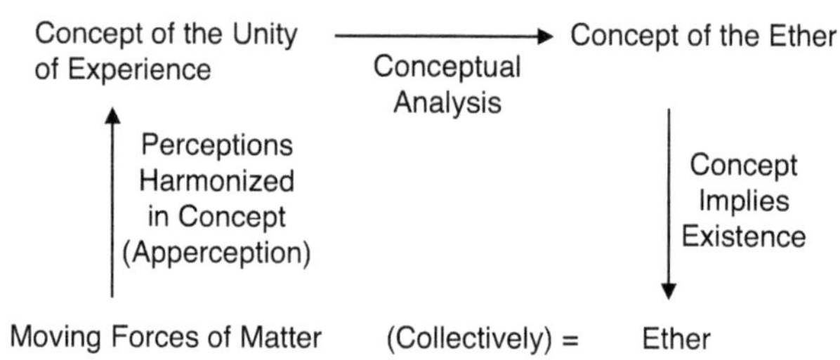

What remains to be explained is *why* this particular relation of concepts takes place. In other words, why does the concept of the unity of experience *contain* the concept of the ether instead of the concept of some other object? In this respect, Kant's Ether Deduction takes on the additional task of being a *demonstration*. Kant must show that the ether is in fact the *only*

reasonable explanation, i.e., a uniquely *necessary* condition for the unity of experience. Before this demonstration is given, however, one can only say:

4. There must be an actual object that is an analytic consequence of the concept of the unity of experience.

Whereas Kant's deduction strategy is transcendental, his demonstration strategy is that of a *reductio ad absurdum*. This is stated in 21:574.14–20, where he says that the proof will proceed indirectly by showing the impossibility of the conclusion's opposite. Kant hopes to derive a contradiction from an indirect proof. The question, however, is what the *opposite* of the ether might be? Kant claims in *Übergang* 12 that the opposition is not *analytically* logical, i.e., P and ¬P, but is rather *synthetically* defined as 'the real opposition of mutually opposing forces' (*OP* 21:582). This comment is as cryptic as it is intriguing. In a literal sense, Kant seems to have in mind the opposition between attractive and repulsive forces within the dynamic plenum.[30] He wants to claim that the opposite (i.e., no dynamic plenum) would be inconsistent with the unity of experience. Within the context of indirect proof, however, one can interpret Kant's argument as positing a *real* opposition between different *types* of counteracting forces. Two different types of force immediately suggest themselves: mechanical and dynamical. Whereas the former type of force is a *relation* between material bodies, the latter type of force is an original *quality* of matter. There is nothing self-contradictory in conjoining these different types of force. As already noted, Kant believes that mechanical forces depend upon dynamic forces. The idea of a dynamic ether would belong to the latter conception of force, whereas the idea of a mechanically related whole of physical bodies would belong to the former. However, if these two conceptions of force are not analytic opposites, how can they be synthetically opposed? At the beginning of 21:572.25–27/573.1–13 and again in 21:574.14–20, Kant suggests that the *opposite* of the ether is empty space, either inside or outside a physical body, as an object of possible experience. The ether, in contrast, brings about the experience of *full* space. One way then of formulating the synthetic opposition that Kant envisions would be to oppose, on the one hand, a mechanical whole of physical bodies containing or separated by empty space, against the ether with all of its stated properties on the other hand. This captures the opposition that Kant describes in his proof and is consistent with his criticisms of atomism throughout *MFNS* and *OP*. This leads to the following theorem introduction, and using the premises already given above one can at the same time formulate the *reductio ad absurdum:*

5. From (4), The actual object that is an analytic consequence of the concept of the unity of experience can be either (Q) the ether with its stated properties, or (synthetic ¬Q) a mechanical whole of physical bodies containing or separated by empty space [Theorem Introduction] (21:572.25–27/573.1–13 and 21:574.14–20).

i. [Assume (5) synthetic ¬Q for *reductio ad absurdum*] a mechanical whole of physical bodies containing or separated by empty space is the actual object that is an analytic consequence of the concept of the unity of experience.
ii. From (1) and (2), everything external, comprehended within the unity of experience, must be capable of affecting the subject.
iii. From (i) and (ii), empty space can affect the subject.
iv. Contradiction: (3) and (iii).

6. (Q) From (i)–(iv), the ether with its stated properties is the actual object that is the analytic consequence of the concept of the unity of experience [Indirect Proof] (21:574.14–20).

The actual object that is an analytic consequence of the subject's concept of the unity of experience is itself, by *definition,* comprehended within the unity of experience. If something external is to be comprehended within the unity of experience, it must be capable of affecting the subject since experience *itself* is predicated upon affection. When something cannot affect the subject, it cannot be an object of possible experience, and cannot be comprehended within the unity of experience. This is Kant's problem with empty space, and as one can see, his whole argument for the ether hinges upon the rejection of empty space as an object of possible experience. The ether serves as a necessary condition for the unity of experience since it rules out the possibility of the experience of empty space.

It seems to me that the above reconstruction is valid, and there is good reason to believe that it is sound. Kant has not only provided a deduction of the ether, but also a demonstration of *why* it is a uniquely necessary condition for the unity of experience. The proof thus serves as a two-part function. (1) It exposes a gap in the Critical philosophy by showing that the cognitive formal transcendental conditions of *CPR* are not jointly sufficient to guarantee the unity of experience. This is a gap in the transcendental part of the metaphysics of nature and is precipitated by the failure of Kant's argument for Substance in the Analogies. (2) It seems to fill this gap by deriving a concept of the ether *a priori* while also proving the existence of the ether. Kant's post-Critical concept of the ether is consistent with his Critical conception of Substance and the ether itself is a transcendental material condition for the unity of experience. At the end of the day, these transcendental conditions are individually necessary, but are they jointly sufficient to guarantee the unity of experience?

3.4 EVALUATING THE SUCCESS OF THE ETHER DEDUCTION

In this final section, I will evaluate the success of the above reconstruction by first dealing with objections that might be leveled against it. Most generally, one might worry that a transcendental existence proof is a *contradictio in adjecto* from the perspective of the Critical philosophy. In particular, one might think that the conclusion of the Ether Deduction runs afoul of Kant's critique of the Ontological Argument.[31] There Kant claims that the existence of an object can never be proven analytically from *mere* concepts. Existence can only be shown through experience, hence synthetically. Again, however, Kant does not begin the Ether Deduction with the mere *concept* of the unity of experience, but rather with the *existence* of the unity of experience as well as the perceptions caused by the moving forces of matter which are then harmonized by apperception into the concept of the unity of experience. Through conceptual analysis, the concept of the ether is shown to be contained within the concept of the unity of experience. Since the unity of experience exists, the ether exists as well. The starting point of the Ether Deduction is synthetic. In this way, Kant's analytic deduction of the ether is not a deduction from *mere* concepts and hence not susceptible to his criticisms of the Ontological Argument. Likewise, Kant's deduction differs from the atomist's derivation of impenetrability from the concept of an atom (discussed in §2.1). Whereas Kant begins the Ether Deduction with the *existence* of the unity of experience (from which he culls the corresponding concept), the atomist must begin simply with the *concept* of an atom. Kant's argument is properly *transcendental* and claims only that the ether is a necessary condition for the unity of experience. Unlike the Ontological Argument, furthermore, the Ether Deduction does not assert the metaphysical reality of something (*viz.* God) understood as existing independently of its relationship to human experience.[32] I will return to this theme again below when distinguishing transcendental philosophy from traditional metaphysics.

One might also be worried that there is a "neglected alternative" to the theorem I introduce in premise five of the Ether Deduction, e.g., couldn't Descartes' material plenum serve just as well to rule out the experience of empty space? Admitting this possibility, it seems as if there are now three options for specifying precisely which kind of actual object could be an analytic consequence of the concept of the unity of experience: (1) A mechanical whole of physical bodies containing or separated by empty space, (2) the ether with its previously mentioned properties, or (3) a Cartesian material plenum.

(1) has already been ruled out by the Ether Deduction itself. Assuming that the actual object, which is an analytic consequence of the concept of the unity of experience, is a mechanical whole of physical bodies containing or separated by empty space leads to a contradiction. Atomistic or corpuscularian views simply cannot safeguard the unity of experience. (2) was my initial conclusion by indirect proof. This leaves us to evaluate the merits of (3). This option seems to me far stronger than Kant's ether thesis. It is also stronger than what could be considered an *analytic* consequence of the concept of the unity of experience. Descartes attempts to rule out not merely the experience of empty space but also the very possibility of empty space. It seems to me, however, that Kant's transcendental deduction of the ether is meant

only to rule out the possibility of the *experience* of empty space, though he remains agnostic with regard to the metaphysical *possibility* of empty space itself. Although the ether is a transcendental condition for experience and as such is transcendentally necessary, it is not metaphysically necessary. In this way, Descartes' thesis is stronger than Kant's thesis, since Descartes attempts to show that something, in some ways similar to the ether, is metaphysically necessary. I will discuss the distinction between transcendental and metaphysical necessity at greater length below.

Of course, one could be a Kantian-minded Cartesian, in which case the material plenum would not be metaphysically but only transcendentally necessary.[33] In this case, I think the dynamist could respond by saying that this is contrary to our experience of the world. As mentioned above, our spatial experience is not of a world "filled up" with matter. Rather, we experience spatially discrete physical objects (or substances). Claiming that the material plenum is only a determinable whereas discrete physical objects in motion are its determinations does not help here insofar as one means by this something like Aristotelian prime matter, which is not at all actual.[34] At the end of the day, the Cartesian seems ill-equipped to deal with the nature of our spatial experience.[35] Insofar as Kant's transcendental arguments are aimed at making sense of the world as we experience it, it seems as if the Kantian-minded Cartesian would have a difficult time being a good Kantian after all.

A closely related worry is that the theorem introduction of premise five is a violation of first-order logic, since the opposition Kant has in mind is not a logical opposition. As Kant says, however, he is not interested in a *logical* opposition but rather a *real* opposition. A deeply important feature of the modern era is the dispute between those who defend a mechanical theory of matter and those who defend a dynamical theory of matter. Even so, Kant believes that mechanism need not be inconsistent with dynamism in the sense that the two theories are not mutually exclusive. Although Kant does not exclude mechanics from his system (it certainly operates at the level of physical bodies), he does want to show that a dynamical theory of matter is the only possible view that can account for the unity of experience. Kant might argue that the theorem introduction in premise five is exhaustive when considered from the transcendental perspective. He would certainly allow that although other kinds of matter are metaphysically or logically possible, they are irrelevant when it comes to answering his particular question, *viz.* is empty space a possible object of experience (mechanics) or is it not (dynamics)? Given the fact that empty space is not an object of possible experience, the rational power of logic compels one to accept a dynamical theory of matter on the assumption that such a theory is the only way of accounting for the impossibility of the experience of empty space in a way consistent with our spatial experience.

This assumption does not seem so farfetched when one realizes that what is most important to Kant's conception of the ether is its *function*. Whatever type of material totality it is that plays the role of ruling out the experience of empty space while preserving all the features of our spatial experience should be called "ether." Correspondingly, any relevant "neglected alternative" to the ether would have to: (1) perform very similar functions to the ones that Kant attributes to the ether, (2) be conceptually inconsistent with Kant's conception of the ether, and (3) exist as an analytic consequence of the concept of the unity of experience.

In the worst case scenario—a scenario I think Kant can avoid—he would not be entitled to a theorem introduction in premise five, but rather only to pose a *dilemma* for explaining the material transcendental condition for experience similar in form to the trilemma he poses for describing space and time in the Transcendental Aesthetic.[36] Instead of $Q \vee \neg Q$, which would be the theorem introduction, Kant could pose the dilemma $Q \vee S$ for describing the material transcendental condition for experience. If the assumption of S, where S = a mechanical whole of physical bodies containing or separated by empty space, leads to a contradiction, then by disjunctive syllogism Q, where Q = the ether with its stated properties. The pressure would now be back on the ether critic to provide a *relevant* "neglected alternative" that meets the above requirements.

As mentioned above, Guyer claims that there are actually four different *types* of Ether Deductions in the *Übergang* section. At different points, Kant argues that the ether is necessary for (1) the unification of space and time, (2) the ruling out of empty space, (3) the transmission of forces required for perception, and (4) the original cause of motion. The above reconstruction seems focused on (2) and one might worry that Kant's argument in *Übergang* 11 fails to establish the other three. Once one connects the Ether Deduction with Kant's discussion of Substance in the Analogies, however, it is easy to see how the ether relates to (1). Once Kant's dynamical theory of affection is fully worked out (see §5.5), it will be clear how the ether relates to (3). Finally, given Kant's claims beginning in *MFNS* and continued throughout *OP* that mechanical forces require dynamic forces, it is easy to see how the ether fulfills the function of (4) as well. Focusing just on (2), however, Guyer believes that such a claim does not follow from Kant's arguments in the Analogies. As he says:

> More problematic, however, is Kant's apparently untroubled inference from the claim that all objects must be uniquely ordered, which is to say positioned, in a single all-encompassing space to the conclusion that an all-pervasive system of matter is required to do this. This is troubling because the third "Analogy of Experience" of the first *Critique* argues that any two objects may be assigned determinate spatial positions relative to each other as long as there are dynamic relations of interaction between them, but does not specify that there must be a continuous bridge of matter between them.[37]

In this quote, Guyer seems to overlook Kant's own claim that there must be "matter everywhere" in order to account for the subject's experience of the simultaneity of objects in the Third Analogy. To put things in Guyer's terms, in order for any two objects to be *assigned* determinate spatial positions relative to each other, there must be matter everywhere making the perception of the subject's position relative to these objects possible. According to Kant, it is only through the 'reciprocal influence' (*CPR* A213/B260) of this material that the subject can establish the simultaneity and consequently the coexistence of objects in space. As I have argued, I believe that this "matter everywhere" is the same as what I have called "Substance," and although Kant's argument for Substance in the Analogies fails, one can find another argument for something that can provide all the same functions of Substance in the Ether Deduction. It is important to remember, however, that Kant's post-Critical theory of Substance takes *force* not *stuff* as fundamental for matter. Consequently, Guyer is on to something insofar as compositional stuff everywhere is not required for the dynamical relations of objects. At the same time, however, he does not recognize that a dynamic force plenum comes to stand in for this compositional stuff in *OP* and is required for the dynamical relations of objects.

Even if we accept Kant's claim in the Third Analogy and in the Ether Deduction that empty space cannot be perceived, Guyer worries that this is merely an *empirical* assertion about the perception of space which is insufficient to ground a *transcendental* deduction of the ether. According to Guyer, transcendental arguments must begin from synthetic *a priori* premises that are neither merely empirical nor simply conceptual. Guyer tries to salvage Kant's argument, however, by appealing to the formal properties of space. Since all regions of space are continuous, there can be no empty spaces which would introduce absolute boundaries between the regions of space. Although Guyer admits that this strategy relies upon the soundness of Kant's arguments in the Transcendental Aesthetic, the argument would at least begin from a premise that is synthetic *a priori*.[38]

If Guyer is right on both scores, his criticisms would pose serious problems not only for my reconstruction of the Ether Deduction, but also for my reconstruction of Kant's arguments in the Analogies. It is unclear, however, whether the starting point of a transcendental argument must be synthetic *a priori*. In the *Prolegomena to Any Future Metaphysics,* Kant makes a distinction between progressive and regressive arguments.[39] In the present context, one could understand the distinction as follows: whereas progressive arguments begin with facts of experience and then infer certain synthetic *a priori* truths as conditions of their possibility, regressive arguments begin with synthetic *a priori* truths and then infer their necessary conditions. In this respect, the Ether Deduction (as well as the arguments in the Analogies) can be considered progressive transcendental arguments. The structure of the Ether Deduction is $P \rightarrow Q$ where one begins with a fact of experience, *viz.* that it is united (P) and derives a necessary condition (Q) for the truth of that claim. Although the starting point of this transcendental argument is contingent, the conclusion of such an argument is synthetic *a priori*. As Kant says at the end of the Analogies, the arguments for the principles of the Analogies 'attempt to prove intellectual and at the same time synthetic *a priori* propositions' (*CPR* A216/B263). In the First Analogy, Kant begins with the subject's experience of simultaneity and succession. In the Ether Deduction, Kant begins with the unity of experience. A synthetic *a priori* proposition is not the starting *premise* of these transcendental arguments, but rather is what that argument aims to *conclude*.

The proof structure of the Ether Deduction also offers a response to Hoppe's dilemma for the argument mentioned in the Introduction (§0.4). Hoppe claims that either (1) the unity of experience simply depends on the combinatory faculty of the understanding in which case the assumption of the ether is trivial and so dispensable, or (2) one must assume a pre-established harmony between our understanding and the ether as a material principle of unity, but Kant rejects this kind of pre-established harmony in *CPR*. In the Ether Deduction, Kant does not think that the combinatorial functions of the understanding are sufficient to guarantee the unity of experience. In fact, the Ether Deduction is predicated on the formal conditions of experience (R) being insufficient for the unity of experience (P). Consequently, Kant must reject Hoppe's first option which forces him onto the second horn of the dilemma. The harmony between our understanding and the ether, however, is not taken to be *pre*-established in the Ether Deduction, but is rather *established* by apperception. As mentioned above, the role of apperception (spontaneous self-affection) in the Ether Deduction is to *harmonize* the perceptions generated by the moving forces of matter within the concept of the unity of experience.[40] Kant will expand on this idea in *Convoluts* 10–11, to which we will turn in the next chapter.

At this point, it is important to carefully measure the success of the Ether Deduction both in terms of what Kant may have *intended* and what he in fact *achieved*. With regard to the latter, I believe that Kant's argument for the actuality of the ether as a material transcendental condition for experience is successful. Kant's Ether Deduction provides both an argument for the totality of empirical reality merely presupposed in the Transcendental Ideal of *CPR* while offering a substitute for Kant's failed argument for Substance in the Analogies. Even so, the Ether Deduction still fails with regard to its stated intention. If we place any value on the title of the section (*Übergang*), one might wonder whether Kant succeeds in effecting a transition from the metaphysical foundations of natural science to physics. As Tuschling make clear (see §0.4), the ether as a collective whole, and as the ultimate object of physics, is itself not an object of experience. It is not empirically knowable, though nevertheless an *a priori* necessary condition for the unity of experience. This is not to say, however, that the ether is itself the object of an intellectual intuition, a noumenal object, or a *metaphysically* necessary object. The ether is *transcendentally* necessary *for* the unity of experience, but is itself not an object *of* experience. The idea that the ether is actual but is not an object of experience is not new to the *Übergang* section. In fact, Kant has held this view of the ether since the *Elementarsystem* where he claims that 'the all-spreading caloric or otherwise named material (ether) is not hypothetical but is also not simply taken from experience (i.e., as belonging to physics)' (*OP* 21:192). This view is reinforced in the *Elementarsystem* by Kant's thoroughly *negative* categorical characterization of the ether relative to the kinds of properties physical bodies possess (see §2.2) and he continues to subscribe to this characterization of the ether in the *Übergang* section.[41] At one point, in *Übergang* 8, Kant claims that the ether 'is not an object of the senses, but rather of sensibility' (*OP* 21:550). Beth Lord has interpreted this passage as claiming that the ether affects the senses beneath the threshold of apperception and recognition.[42] It is the 'dynamical horizon within which bodies arise and move.'[43] I agree with Lord on this latter point and have argued that the categories only have legitimate application to substances (in contrast to the ether/Substance) in empirical intuition (see §1.3). As I will argue below (§4.1 and §5.5), even though we do not have empirical cognition (objective perception) of the ether, being affected by the ether in sensibility (subjective perception) is a necessary condition for the subject's empirical cognition of substances.

It is sometimes difficult, admittedly, to distinguish metaphysical theses from transcendental theses in *OP*. This leads Westphal to wonder: 'What exactly is transcendental philosophy and what distinguishes it from metaphysics.'[44] Westphal believes that Kant's transcendental arguments actually entail strong metaphysical conclusions that undermine his transcendental idealism. On pain of subjective idealism, he argues, Kant must admit the existence of a transcendental *material* condition of experience. According to Westphal, however, such an admission undermines Kant's transcendental idealism since the latter allows only for transcendental *formal* conditions of experience, consequently prohibiting the cognition of objects that are transcendentally real.[45] This is an important objection to my thesis since, if Westphal is right, the addition of the ether as a transcendental material condition of experience in the Ether Deduction would itself undermine Kant's Critical philosophy under the assumption that the ether is transcendentally real, i.e., a thing-in-itself or noumenal object. I believe, however, that there is another way of understanding Kant's transcendental material condition of experience.

First, it is important to note that Kant himself says that the ether is not a thing-in-itself or noumenal object, but rather a phenomenon, given its affective relations to the subject. As he says near the beginning of *Convolut* 10, written immediately after the *Übergang* section, 'The first principle of representation of the moving forces of matter [is] to regard them not as things-in-themselves, but rather as phenomena according to the relations which they have to the subject.'[46] Although Kant sometimes refers to the ether as the 'thing itself [*Sache selbst*],' he is usually careful to distinguish this concept from the concept of a noumenon.[47] It must be admitted, however, that these are murky waters. How can something be a phenomenon without being an object of experience? Like the notion of a transcendental material condition, one might think that this idea is itself a *contradictio in adjecto* from the perspective of the Critical philosophy. Although I believe that Kant can overcome this worry, it requires him to draw a distinction that he does not draw in the *Übergang* section, *viz.* between direct and indirect appearances. This is a distinction that is first drawn in *Convolut* 10 (written immediately after the *Übergang* section) and that he continues to make use of in later parts of *OP*.[48] As I will argue below (§4.1 and §5.5), whereas the ether *directly* appears to the subject, objects of experience *indirectly* appear and depend upon the activity of the subject organizing direct appearances.[49] In *MFNS,* Kant claims that 'matter is the moveable insofar as it, as such a thing, can be an object of experience' (*MFNS* 4:554). Since the ether (or Substance) is not locomotive, it cannot be an object of experience. Physical bodies (or substances), however, are locomotive and so can be objects of experience. The latter are indirect appearances which themselves depend upon the direct appearances of the ether.

Returning to Westphal's question, however, I would answer that transcendental philosophy is concerned with establishing the conditions for the possibility of human *experience* and the *unity* of that experience, whereas the metaphysical speculation of transcendental realism need not make any appeal to the experience of creatures like us when establishing its conclusions. Although Kant is concerned, throughout the Critical and post-Critical periods, in developing a metaphysics of nature, this metaphysics cannot be understood independently of its transcendental foundation. In fact, as I have argued, Kant hopes to complete this transcendental foundation (i.e., the transcendental part of the metaphysics of nature) through his post-Critical transition project. Unlike dogmatic metaphysics, however, in transcendental philosophy human experience is the tether to which metaphysics is bound. Kant makes this point clearly in *MFNS:*

> Undoubtedly they [natural philosophers] have understood by the latter [metaphysics] the folly of contriving possibilities at will and playing with concepts, which can perhaps not be presented in intuition at all, and have no other certification of their objective reality than that they merely do not contradict themselves. All true metaphysics is drawn from the essence of the faculty of thinking itself, and is in no way fictitiously invented on account of not being borrowed from experience. Rather, it contains the pure actions of thought and thus *a priori* concepts and principles, which first bring the manifold of *empirical representations* into the law-governed connection through which it can become *empirical* **cognition,** that, is experience.
>
> (*MFNS* 4:472)

Going back to *CPR,* this constitutes the heart of Kant's critique of transcendental realism (i.e., early modern Rationalism) as a philosophical position. The metaphysical speculation of transcendental realism has no place in transcendental philosophy insofar as transcendental realism concerns itself with establishing the actuality of objects which may bear no connection 'with the material conditions of experience' (*CPR* A218/B266), i.e., with the material conditions of conscious experience for creatures like us.[50] In contrast to these forms of metaphysical speculation, Kant's Ether Deduction not only establishes the *actuality* of an object that bears a connection with the material conditions of experience, but also one whose connection with the actual 'is determined in accordance with the general conditions of experience.'[51] Hence, the ether is not only actual, but in accordance with the Third Postulate of Empirical Thought, also materially *necessary.*[52] There is, consequently, an important difference between the way Kant understands the actuality of the ether and the way that transcendental realism understands actuality. Whereas transcendental philosophy must always bear in mind the limitations placed upon it by the Postulates of Empirical Thought, classic metaphysics need not recognize any such limitations. One can see this latter type of thinking exemplified in Descartes' theory of a material plenum as well as Leibniz's theory of monads. In the *Übergang* section, Kant emphasizes the material necessity of the ether, calling it a 'necessary phenomenon':

> The attributes of this [material] (since it is all-embracing, *individual* (*unica*) and the basis of all [forces] for the unity of the object of the one experience) are given according to the principle of identity: namely, that it is *universally distributed, all-penetrating,* and *all-moving* (not that it is itself movable (*locomotiva,* that is, displaceable)). And as such, it is necessary, that is, *permanent.* For *sempiternitas est necessitas phaenomenon* (*OP* 21:584).[53]

Kant's post-Critical conception of the ether's modality seems perfectly consistent with his Critical conception of material necessity in the Third Postulate. As a transcendental material condition of experience, the ether is both transcendentally and materially necessary. Whereas the formal conditions of experience (space, time, categories, and apperception) are transcendentally necessary insofar as they are *required* for the possibility of experience for creatures like us, the ether is also materially necessary in addition to being transcendentally necessary since its *actuality* is required as a condition of possible experience. As Kant says in *Übergang* 8, 'the all-penetrating caloric is the first condition of the possibility of all outer experience' (*OP* 21:551).

Updating Kant's conception of transcendental material necessity, the necessity of the ether can be understood in the following way: in all and only possible worlds where there is a unity of experience and creatures cognitively similar enough to us, the ether exists. Transcendental material necessity is best understood as a kind of *restricted* nomological necessity. Beyond the cognitive restriction, transcendentally possible worlds involve certain *natural* restrictions as well given the fact that there is a unity of experience in these worlds. What *experience* is must be held constant. The natural laws of transcendentally possible worlds must be similar enough to the actual world's natural laws such that affective relations between spatiotemporal objects and receptive subjects still obtain. There must be objects in these worlds causally-nomologically similar enough to the objects in the actual world so that these objects are capable of affecting the subject in the same basic ways as in the actual world. As Kant says throughout *CPR, outer* experience of actual objects is a necessary condition for experience *überhaupt*.[54] Outer experience in Kant's sense requires that actual objects affect receptive subjects in sensibility.

I have described the ether as transcendentally necessary in order to emphasize that it enjoys the same modal status as the other transcendentally necessary *formal* conditions of experience, *viz.* space, time, categories, and apperception. As mentioned above, however, the transcendental necessity of the ether in particular, corresponds closely to Kant's conception of "material necessity" in the Third Postulate. It is important to note that Kant also explicitly distinguishes material necessity from logical or formal necessity in the Third Postulate.[55] As one should see at this point, Kant's conception of the ether does not commit him to transcendental realism and its claims of metaphysical necessity. Transcendental material necessity is simply a far more circumspect modal position whose proper subject is the conditions for the possibility of experience determined in accordance with the Postulates of Empirical Thought. In contrast, the metaphysical speculation of transcendental realism is neither constrained by the Postulates of Empirical Thought nor must the conditions of conscious experience for creatures like us play a constitutive role in this speculation. The conclusion of the Ether Deduction establishes only the transcendental material necessity of the ether. This is modally weaker than the metaphysical necessity of the transcendental realist position that Westphal attributes to Kant. Since the ether is actual only relative to the *conditions of possible experience,* it is transcendentally ideal. Since the ether is *actual,* according to these conditions, it is likewise empirically real.[56] If I am right, then my interpretation would be preferable to Westphal's view according to my second principle of interpretation (see §0.1) since it makes Kant consistent with himself.

Although the ether cannot be an object of experience, it is nonetheless *actual* insofar as it is the transcendentally necessary material condition of experience. Furthermore, it serves as the material condition for the all the actual objects of experience. As mentioned above, the great irony of this conclusion is one that Tuschling is quick to point out, and one to which I have already alluded. If the Ether Deduction effects the transition from the metaphysical foundations of natural science to physics, it must show that the ether is the ultimate object of physics. Although Kant is able to prove the existence of the ether, he is unable to prove that it is an object of physics, understood as an object of experience, on pain of undermining the very transcendental proof strategy of the Ether Deduction.[57] If the ether were an object of experience, it could *in principle* be cognized *a posteriori.* If the ether could be cognized *a posteriori,* however, it could not serve as the transcendental material condition of experience, since the transcendental conditions of experience must be ascertained *a priori.* It is difficult to see how the ether could be cognized *a posteriori* in any case since the category of substance cannot apply to the ether. The ether is simply Kant's post-Critical conception of Substance. Much as the category of substance cannot apply to Substance in the Analogies (see §1.2), so too does it fail to apply to the ether in *OP.*

For much the same reason, the transcendental conditions of experience, as they stand at the end of the Ether Deduction, are insufficient for establishing the unity of experience. While the formal transcendental conditions of experience and the material transcendental condition of experience are individually necessary for the unity of experience, Kant cannot establish their joint sufficiency since he has no way, in the Ether Deduction, of showing how the categories (transcendental formal condition) apply to the ether (transcendental material condition), as the ultimate object of physics. Without knowing if the conceptual conditions of experience even apply to the material condition of experience, Kant has no way of precluding some *in principle* discontinuity between them. My discussion of *Convoluts* 10–11 in the next chapter will center on how Kant can rule out this possibility even if the categories do not have direct application to the ether. As we will see, Kant broadens his definition of "physics" in *Convoluts* 10–11, which helps to clarify how the ether could be the ultimate object of physics even if it is not an object of experience as physical bodies are. The key to ruling out the possibility of discontinuity between the formal and material conditions of experience, however, is the *a priori* concept of the ether, which Kant has *derived* in the Ether Deduction, but has not yet *integrated* with the formal conditions of experience. Although Kant does not do this in the Ether Deduction, it is something he does in *Convoluts* 10–11. Only by understanding the role of the ether both as a formal and material condition of experience can Kant finally bridge the gap in his Critical philosophy.

From an argumentative standpoint, one can view the Ether Deduction as completing Kant's argument for the *necessary* conditions for the unity of experience, i.e., $P \rightarrow (Q \wedge R)$.[58] In *Convoluts* 10–11, however, Kant provides an argument for the *sufficient* conditions for the unity of experience, i.e., $(Q \wedge R) \rightarrow P$. Instead of focusing on the ether as a transcendental material condition for the unity of experience, however, Kant will direct his attention to the role that the *concept* of the ether plays as a transcendental formal condition for the unity of experience. Kant will argue that the transcendental formal conditions in conjunction with the transcendental material condition of experience are sufficient to insure the unity of experience. Consequently, the Ether Deduction and *Convoluts* 10–11 together constitute Kant's argument for $P \leftrightarrow (Q \wedge R)$.

One can view Kant's Ether Deduction in one respect as an achievement, but in another respect as a failure. Although the Ether Deduction adds a transcendental material condition which, along with the transcendental formal conditions of experience, is necessary for the unity of experience, the Ether Deduction fails to show that the formal transcendental conditions of experience are jointly sufficient with the material transcendental condition of experience for the unity of experience. Without establishing the latter, Kant cannot effect a transition from the metaphysical foundations of natural science to physics and consequently cannot fill the gap that lies at the very heart of his Critical philosophy. Nevertheless, one cannot evaluate the Ether Deduction simply on the basis of what Kant may have intended, but one must also take into account what Kant does in fact *achieve*. Insofar as Kant is able to establish the existence of the ether as the *material* transcendental condition of experience and derive a new *a priori* concept of the ether or Substance, the Ether Deduction should be considered a major post-Critical development of Kant's Critical philosophy.

NOTES

1. Compare the section I have chosen to translate (*OP* 21:572–573) with the amanuensis copy translated in the Cambridge edition (*OP* 22:550–551). Although our translations differ to some degree, the original German is nearly identical.
2. Ernst-Otto Onnasch, "Der Briefwechsel zwischen Immanuel Kant und Jeronimo de Bosch: Oder ein Beitrag zum holländisch-deutschen Austausch über die kritische Philosophie," *Kant-Studien* 102, no. 1 (2011): 89–112 (original Latin and German translation on pages 104–108). Onnasch refers to the letter in question as *Leiden 2* (original Latin and German translation on pages 104–108).
3. There is a discrepancy between Rink's date (August 14th, 1799) and Kant's date (July 20th, 1799).
4. Here I am following Onnasch's rendering of the Latin "*concinnando*" within the context of the letter. Although he translates this into German as "*Komposition,*" he argues that this refers to a later stage of the writing process where Kant would be compiling his work for editing. As we know, some of these drafts would soon be found in amanuensis copies. Onnasch also argues, as I do, that Kant's transition project from this period should be distinguished from earlier drafts where Kant is still working out leftover problems from *MFNS*. See Onnasch, "Der Briefwechsel zwischen Immanuel Kant und Jeronimo de Bosch," 96–97.
5. Förster notes that Kant did cross out a section of the amanuensis copy of the Ether Deduction at some point (*viz. OP* 22:552–555). See Förster, *Kant's Final Synthesis,* 88. Apparently, Kant did not, however, strike out the section of the amanuensis copy that is almost identical to the section that I translate.
6. Bold emphasis added. As I argued in §2.0, Kant uses the terms "ether" [*Äther*] and "caloric" [*Wärmestoff*] interchangeably in the *Übergang* section.
7. Förster, *Kant's Final Synthesis,* 89–101. For a shorter reconstruction of the Ether Deduction which also does not focus on a particular draft, see Emundts, *Kants Übergangskonzeption,* 181–201, especially 181–182.
8. Guyer, "Kant's Ether Deduction and the Possibility of Experience," in *Kant's System of Nature and Freedom* (Oxford: Oxford University Press, 2005), 76.

Notwithstanding the different lines of argumentation, Guyer recognizes (as I do) that Kant is arguing for the actuality of the ether as a transcendental condition of experience.

9. Wing-Chun Wong also connects the Ether Deduction back to the Analogies of Experience. He argues that the Analogies construct the temporal aspect of the ether (sempiternal) whereas the Ether Deduction constructs the spatial aspect of the ether (omnipresent). He does not, however, recognize that the Third Analogy already anticipates the spatial aspect of Kant's argument in the Ether Deduction. See Wing-Chun Wong, "On the Idea of an Ether-Deduction in the *Opus postumum*," in *Kant und die Berliner Aufklärung. Akten des IX. Internationalen Kant-Kongresses,* eds. Volker Gerhardt, Rolf-Peter Horstmann, and Ralph Schumacher (Berlin: De Gruyter, 2001), vol. 4, 679–680.
10. For example, see the quote that began the last chapter.
11. All translations from *Übergang* 11 are my own.
12. See *CPR* B145.
13. There is a sense in which Kant's post-Critical formulation is stronger than what he would allow for in the Critical period. Whereas in the Critical period Kant believes that the 'collective unity of the whole of possible experience' is a 'necessary problem for reason' (*Prolegomena,* 4:328), in the post-Critical period this serves as the starting point for the Ether Deduction and Kant aims to establish the actuality of the ether as a necessary condition for this stronger form of unity. As we will see in Chapter Four, Kant ceases to think that a distributive unity properly describes what he means by "experience." In any case, my reconstruction of the Ether Deduction will endeavor to show how the ether is necessary not only for Kant's post-Critical conception of the "unity of experience," but for his Critical conception as well.
14. One might wonder whether Kant is returning to a certain kind of pre-Critical transcendental realism post-Critically in *Übergang* 11. As mentioned in the previous chapter (§2.1), in the *New Elucidation* Kant uses God constitutively in a way that seems strikingly similar to his use of the totality of empirical reality in the Transcendental Ideal. I have argued that Substance or ether is this totality of empirical reality. If Kant's use of God in the *New Elucidation* commits him to transcendental realism, isn't he equally committed to transcendental realism in *Übergang* 11? I think this continuity is useful insofar as it illustrates the fact that the ether is not really a new idea for Kant and has been there in one guise or another throughout his philosophical career. Granting such continuity, I would want to claim, nevertheless, that Kant's post-Critical Ether Deduction is the only sound argument for a pre-Critical object that was improperly used in the Critical era. Until the *Übergang* section, Kant lacked an argument for this object that would be consistent with his Critical philosophy, and *this* is what commits Kant to transcendental realism in the *New Elucidation.* Kant's problem lies with his justification for the *actuality* of the object, not the *function* the object has within the pre-Critical or Critical philosophy which remains unchanged (in some respects) post-Critically.
15. Translation is mine.
16. For further discussion of how Kant uses the principle of thoroughgoing determination in *OP,* see Rohs, "Kants Prinzip der durchgängigen Bestimmung alles Seienden," 177–180 and So-In Choi, *Selbstbewußtsein und Selbstanschauung: eine Reflexion über Einheit und Entzweiung des Subjekts in Kants Opus Postumum* (Berlin: De Gruyter, 1996) 120–125.
17. See *MFNS* 4:476.
18. See *MFNS* 4:536.
19. At this point, the mechanist might object that Kant is begging the question since he is assuming the very dynamic model in premise (2) that he aims to

prove through the Ether Deduction. As we will see in premise (5) and following, however, Kant provides good reasons (that do not depend on his dynamic conception of affection) for thinking that the mechanical model is incapable of insuring the unity of experience.

20. Kant already connects apperception with self-affection in the Critical period. See *CPR* B153–156. The relationship between the moving forces of matter and apperception will be discussed at greater length in §4.3.
21. See Edwards, *Substance, Force, and the Possibility of Knowledge,* 147–166. Beiser also endorses this view. See Beiser, *German Idealism,* 635n9.
22. Guyer, "Kant's Ether Deduction and the Possibility of Experience," 82.
23. See also my earlier discussions of this issue (§1.1, 1.3, and 2.1).
24. For the Second Postulate, see *CPR* A225/B272.
25. Here I am using the term "object" in its general signification corresponding to Kant's use of the term "*Objekt*" which is often distinguished from a particular empirical object or "*Gegenstand.*" I will argue in §4.1 that although subjects have objective perceptions (cognitions) of substances, they can only have subjective perceptions (sensations) of the ether.
26. Translation is mine.
27. Translation is mine. See also Alexander Rueger, "Brain Water, the Ether, and the Art of Constructing Systems," *Kant-Studien* 86, no. 1 (1995): 37–38. There Rueger spends some time discussing the derivation of the ether concept from the concept of the unity of experience. Like me, Rueger believes that the ether concept must be *a priori* and not empirical if it is to fulfill its transcendental function within the context of *OP.* In contrast to my position, however, Rueger does not think there is a material analogue to the ether concept. Much like Förster, he believes the ether is merely an ideal of reason. His fear is that by adding a genuine transcendental *material* condition for experience Kant might undermine his deductions of the transcendental *formal* conditions of experience in *CPR*. Rueger is not the only commentator on *OP* to worry about the implications of adding a transcendental material condition of experience. Adickes himself pointed out the worry that Rueger hopes to avoid. See Adickes, *Kants Opus postumum,* 393–395. Choi even goes so far as to spend a whole book discussing the role that the transcendental formal conditions of experience play in *OP* while completely ignoring the role that the ether might play as a transcendental material condition of experience. See Choi, *Selbstbewußtsein und Selbstanschauung.* In contrast, Edwards and Schönfeld insist upon the existence of the ether as a genuine transcendental material condition of experience, and even recognize the potential conflict between the transcendental formal and material conditions of experience, but do little to resolve the conflict beyond saying that the price of removing the transcendental material condition of experience from Kant's considered view (which requires ignoring the Third Analogy and isolating Kant's pre-Critical and post-Critical philosophy of nature from the Critical core of *CPR*) is higher than the price of leaving it in (potential tension between the transcendental formal and material conditions of experience in *CPR*). See Edwards and Schönfeld, "Kant's Material Dynamics and the Field View of Physical Reality," 113–114. Lucia Procuranti is one of the few commentators to distinguish between the *a priori* concept of the ether and the actual extension of that concept. See Lucia Procuranti, *Il problema della costituzione della materia nella filosofia di Immanuel Kant* (Trient: Pubblicazioni di *Verifiche,* 2004) 224–227 and 253–262. For a discussion of her book (in German), see Basile, *Kants Opus postumum,* 295–296. In what follows, I hope to establish that Kant's considered view requires both the existence of the ether as a transcendental *material* condition of experience as well as the ether concept as one of the transcendental *formal* conditions of

experience. At the same time, I hope to show that Kant can maintain this view without undermining any of the central claims of *CPR*.

28. Although Kant says that the concept of the ether is empirical in this same passage, in the next chapter (§4.2), I will explain why this concept cannot be an "empirical concept" in the sense that Kant uses the term in *CPR*. What I think he means by "empirical" in this passage from *Übergang* 11 is that the concept of the ether is the concept of something *actual*. We cannot have 'empirical consciousness of its principle' (*OP* 21:572.25–27/573.1–13), however, which suggests that it is an *a priori* concept.
29. In the previous paragraph, Kant mentions the *Elementarsystem* and describes the ether as 'thoroughly distributed in cosmic space' as well as 'the basis of all moving forces of matter' (*OP* 21:572.3–15). Translation is mine. I agree with Emundts that Kant's conception of the ether in the *Übergang* section is consistent with his conception of the ether in the *Elementarsystem*. See Emundts, *Kants Übergangskonzeption*, 148.
30. In *CPR*, Kant contrasts logical opposition with real opposition. The latter occurs when 'one reality, if combined in one subject with another, cancels out the effects of the latter' (*CPR* A273/B329). He explains this cancellation in terms of mechanical forces operating in opposite directions.
31. See *CPR* A592–602/B620–630. Hoppe raises this worry. See Hoppe, *Kants Theorie der Physik*, 106–109.
32. Beiser makes a similar point. See Beiser, *German Idealism*, 190.
33. This strikes me as being close to the third alternative argument for Substance that I considered but ultimately rejected in §1.3.
34. Although I will argue in §5.5 that the ether is an undetermined phenomenal object from which determined phenomenal objects emerge, it is still actual and not a mere determinable that becomes actual only through its determinations.
35. Descartes also requires that God constantly conserve the motion in the world. Kant's conception of an internally moving and sempiternal ether, however, obviates the need for such external metaphysical interference. See René Descartes, *Principles of Philosophy*, in *The Philosophical Writings of Descartes, trans. John Cottingham, Robert Stoothoff, and Dugald Murdoch* (Cambridge, Cambridge University Press, 1985), vol. 2, 240 (II:36).
36. See *CPR* A23/B37–38.
37. Guyer, "Kant's Ether Deduction and the Possibility of Experience," 79–80.
38. See Guyer, "Kant's Ether Deduction and the Possibility of Experience," 81.
39. See Kant, *Prolegomena* 2:277. For further explanation of the distinction, see Georges Dicker, *Kant's Theory of Knowledge: An Analytical Introduction* (Oxford: Oxford University Press, 2004), 24–25.
40. Kemp Smith would have a similar worry with my view since it relies only on empirical affection in the phenomenal world. He believes that any view that relies only on empirical affection must assume a kind of pre-established harmony wherein the *a priori* contributions of the subject to experience are adjusted so that they can be relied upon to reconstruct what is independently real. Again, this harmony is not assumed as *pre*-established by Kant but *established* by apperception in the Ether Deduction and *Convoluts* 10–11. See Kemp Smith, *A Commentary to Kant's Critique*, 634.
41. See *OP* 22:610.
42. Lord, *Kant and Spinozism*, 165.
43. Lord, *Kant and Spinozism*, 166.
44. Westphal, "Does Kant's *Metaphysical Foundations of Natural Science* Fill a Gap in the *Critique of Pure Reason?*," *Synthese* 103, (1995): 83n121.
45. See Westphal, *Kant's Transcendental Proof of Realism*, 110–116.
46. *OP* 22:300.

47. When I use "noumenon" in the above context, I am referring to a *positive* noumenon or a metaphysically real object with intrinsic non-relational properties. See *OP* 22:390 for where Kant refers to the ether as the "thing itself" but distinguishes it from a noumenon. At the same time, however, see *OP* 22:471, where Kant refers to the ether as a "thing itself" but does not distinguish it from a noumenon. I will return to this issue in the final chapter.
48. See *OP* 22:320.
49. My interpretation of the distinction is the minority position. For views similar to my own, see Kurt Hübner, "Leib und Erfahrung in Kants Opus postumum," *Zeitschrift für philosophische Forschung* 7, no. 2 (1953): 210–211 and Walter Bröcker, "Das Höhlenfeuer und die Erscheinung von der Erscheinung," in *Die Gegenwart der Griechen im neueren Denken. Festschrift für Hans-Georg Gadamer zum 60. Geburtstag,* eds. Dieter Heinrich, Walter Schulz, and Karl-Heinz Volkmann-Schluck (Tübingen: Siebeck, 1960), 36–37. Beiser uncovers four different versions of the distinction. In addition to the distinction between (1) direct and indirect appearances, there is the distinction between (2) appearance and appearance of an appearance, (3) metaphysical and physical appearances, as well as (4) subjective and objective appearances. See Beiser, *German Idealism,* 201. Like Beiser, I believe that the distinction between direct and indirect appearances can be mapped onto the distinction between appearance and appearance of an appearance. See Beiser, *German Idealism,* 208 and *OP* 22:340. Unlike Beiser (who represents the majority position), however, I do not think that "appearance of an appearance" denotes the ether. When it comes to the distinction between metaphysical and physical appearance, Kant sometimes holds that metaphysical = direct whereas physical = indirect (*OP* 22:325) but elsewhere holds that metaphysical = indirect whereas physical = direct (*OP* 22:340). The fourth distinction seems to reflect the Critical distinction between the form (subjective) and matter (objective) of appearance but where the matter of appearance is understood as the result of the ether's activity. Compare *CPR* A20/B34 with *OP* 22:378. For his part, Mathieu articulates eighteen different meanings of "appearance of an appearance." See Mathieu, *Kants Opus postumum* 144–145.
50. This is from the Second Postulate of Empirical Thought. Emundts also appeals to the Second Postulate when arguing for the reality of the ether as a subjective principle of experience. See Emundts, *Kants Übergangskonzeption,* 163.
51. See the Third Postulate of Empirical Thought, *CPR* A218/B266. Although Kant says that the Third Postulate 'does not hold of the existence of things, as substances, since these can never be regarded as empirical effects' (A227/B280), because what I have called "substances" are *effects* of the ether according to Kant's post-Critical view, it seems as if we can cognize the necessity of the ether as well. Since he does not distinguish between Substance and substances, Edwards argues that the material necessity of the ether (or Substance) would violate the Third Postulate. See Edwards, "Spinozism, Freedom, and Transcendental Dynamics," 60.
52. Bernard Rousset likewise emphasizes that the Ether Deduction in *OP* is consistent with the second and third postulates in *CPR*. See Roussett, Bernard, *La doctrine kantienne de l'objectivité. L'autonomie comme devoir et devenir* (Paris: Vrin, 1967), 135–138. For a discussion of this book (in German), see Basile, *Kants Opus postumum,* 172–181.
53. Note that Kant ties the necessity of the ether to it being "permanent" [*alldaurend*] (in contrast to "persistent" [*beharrlich*]), which reinforces the connection between the ether and Substance. Adickes also takes some time to discuss the modal status of the ether. See Adickes, *Kants Opus postumum,* 400, 459, and 581.

54. See especially the opening moves of the B-edition Introduction (*CPR* B1), Refutation of Idealism (*CPR* B274–279), and General Note on the System of Principles (*CPR* B288–294). Although Kant's post-Critical theory of affection is consistent with what I say here, as we will see in the final chapter, it is significantly more complex than the current discussion suggests.
55. See *CPR* A226/B279.
56. In this respect, I am in agreement with Beiser who holds that Kant's views in *OP* are generally consistent with his views in *CPR*. The Ether Deduction is itself best understood as expanding what can be proven within the context of transcendental idealism rather than committing Kant to a form of transcendental realism. See Beiser, *German Idealism,* 214 and 636n16. For a summary of views that likewise insist upon the empirical reality of the ether, see Basile, *Kants Opus postumum,* 405–409.
57. Tuschling, *Metaphysische und Transzendentale Dynamik,* 175–178. Citing passages from *OP* that were written prior to both the *Elementarsystem* and *Übergang* sections, Beiser argues that the existence of the original moving forces of matter can only be confirmed *a posteriori.* See Beiser, *German Idealism,* 185 and 199. In my view, this does not take into account how Kant's view changed over time and would undermine his transcendental proof strategy in the *Übergang* section.
58. As mentioned above (§3.2), P = the unity of experience, R = the transcendental formal conditions of experience familiar from *CPR,* and Q = the ether.

4 Kant's Transition Project in *Convoluts* 10–11

> 1) What is physics? It is the aggregate of empirical cognition of the moving forces of matter in experience.
> 2) What is experience? It is the unification of perceptions, under the principle of their connection, according to concepts, to a doctrinal system (a subjective whole of the aggregate of perceptions).
> 3) What is the transition from the metaphysical foundations of natural science to physics? It is the doctrinal system of experience in general, applied to the natural system.
> 4) How is the transition from one to the other *possible*: (1) in respect of the material element of the object, (2) of the formal element of the subject?
>
> —Kant, from *Convolut* 11 (*OP* 22:480)

4.0 INTRODUCTION

In the above passage, Kant asks four questions whose answers are absolutely central to understanding the next phase of his transition project. Kant's brief answers offer a snapshot of his thinking in *Convoluts* 10–11, a snapshot that this chapter will aim to explain in detail. There seems to be a close connection between physics and experience where the transition aims to take what is merely an *aggregate* (perceptions generated by moving forces) and transform it into a *system*. In order to effect this transition, one must examine not only the *material* conditions of experience, but the *formal* conditions as well.

Kant saw a need in *Convoluts* 10–11 to revise the way that the transcendental *formal* conditions of experience (*viz.* space, time, categories, and apperception) function empirically given the addition of a transcendental *material* condition of experience in the Ether Deduction. The first section of this chapter examines the revisions Kant makes to the Transcendental Aesthetic (space and time) of *CPR*. The second section examines what changes Kant makes to the Transcendental Analytic (categories) of *CPR*. The third section discusses the new role that apperception plays in *Convoluts* 10–11 and how it helps to solve some important problems with the Critical philosophy. The final section argues that these revisions are necessary for Kant to finally effect a transition from the metaphysical foundations of natural science to physics and that this transition is tantamount to bridging the gap in his Critical philosophy.

4.1 SPACE AND TIME IN THE TRANSCENDENTAL AESTHETIC

As should be familiar from the Transcendental Aesthetic of *CPR*, Kant believes that space and time are *nothing but* subjective forms of intuition *a priori*. They are not things in themselves (Newton) nor are they properties of or relations between things in themselves (Leibniz). Kant claims that space and time are *both* transcendentally ideal and empirically real. Space and time are transcendentally ideal insofar as they are nothing but *a priori* forms of the subject's sensibility and are nothing at all once one abandons the conditions of the subject's sensibility. The purity of space (and by extension time) is directly related to its transcendental ideality. This pure representation of space is, furthermore, presented in its formal aspect as an 'infinite given magnitude' (*CPR* A25/B39–40). This aspect of space, as a pure form of intuition, can be represented to the mind prior to sensation (*viz.* geometrical space).[1] Nonetheless, space and time are also empirically real with regard to all the empirical objects that can present themselves to sensibility. As Kant says with respect to space in the Transcendental Aesthetic:

> Our expositions accordingly teach the reality (i.e., the objective validity) of space in view of everything that can come before us externally as an object, but at the same time the ideality of space in regard to things when they are considered in themselves through reason, i.e., without taking account of the constitution of our sensibility.
>
> (*CPR* A28/B44)[2]

Beginning with the two Metaphysical Expositions of space and time in the Transcendental Aesthetic, Kant makes clear that he does not consider space and time *themselves* to be objects of perception *a posteriori*.[3] That time (and by extension space) 'cannot be perceived by itself' (*CPR* B225) is also important for Kant's arguments in the Analogies.[4] At the same time, however, a 'substratum must be encountered that represents time in general' (*CPR* B225). As I argued in Chapter One (§1.1), Substance is supposed to fill this role and insures the empirical unity of time. Likewise, when it comes to space, Kant has long claimed that space must be filled by Substance (Third Analogy) or by moving forces (Ether Deduction) in order to rule out the possibility of the experience of empty space since this would undermine the empirical unity of our spatial experience. In the quote from *Übergang* 2 that began Chapter Three, Kant even says that the ether is 'hypostatized space itself.' In *Convolut* 11, Kant echoes this idea when he says that 'matter is what makes space into an object of the senses' (*OP* 22:508). As I argued above (§2.2), in *OP* matter is fundamentally force. In the present context, the way to understand "matter" is in terms of the original moving forces of attraction and repulsion which are not themselves objects of experience but are necessary for our experience of empirical space. Kant makes this point in *Convolut* 11 when contrasting his position with Newton's view:

> To this Newtonian principle of universal attraction through empty space there corresponds a similar principle of repulsion (*virium repelentium*), which, likewise, cannot be an object of experience in itself, but is only necessary in order to present space as an object of the senses.
>
> (*OP* 22:529–530)[5]

In *Convolut* 10, Kant connects this conception of the ether back to the *Elementarsystem* (discussed in §2.2): 'The moving forces of matter, however, in virtue of the unity of space and its thoroughgoing fullness (since empty space would be no object of experience), form an *elementary system,* which is, indeed, the object of physics' (*OP* 22:342). Although Kant must prove that the ether is the ultimate object of physics through his transition project, it cannot be an object of *experience* given his wholly *negative* characterization of the ether in the *Elementarsystem* (relative to the properties that substances possess) and the *a priori* proof strategy of the Ether Deduction.

The way I think we should interpret Kant's view in *Convoluts* 10–11 is to say that the omnipresent moving forces of the ether are necessary to make space and time thoroughly *perceptible,* though this does not insure there is any *particular* thing that is *perceived.* As mentioned in the last chapter (§3.3), the dynamic ether or Substance that fills space *precludes* the experience of empty space, even if it does not *include* the experience of anything in particular, i.e., substances. In *Convolut* 10, Kant distinguishes between two ways in which things appear in space and time:

> The appearance of things in space (and time), however, is two-fold: (1) that of objects which we ourselves insert in space (*a priori*), and which is metaphysical; (2) that which is empirically given to us (*a posteriori*), and which is physical. The latter is direct appearance, the former indirect—that is, appearance of an appearance. The object of an indirect appearance is the thing [*Sache*] itself—that is, one which we only extract from intuition, insofar as we ourselves have inserted the appearance, that is, insofar as it is our own cognitive product. . . . We made the object of empirical intuition (of perception) ourselves; produced it ourselves for the instrument of sensation (by composition); and thus presented a sense-object for experience in accordance with the latter's universal principles.
>
> (*OP* 22:340–341)

Here Kant makes a distinction between "direct appearance" and "indirect appearance" (or "appearance of an appearance"). Although direct appearances are *given* to the subject, indirect appearances are objects *produced* by the subject. Does this mean that indirect appearances are created out of whole cloth? What is the relationship between direct and indirect appearance? The key, I believe, is the concept of an "appearance of an appearance" which is identical to "indirect appearance." My suggestion is that direct appearances are of the moving forces of matter affecting the subject. Earlier in *Convolut* 10, Kant says that 'experience does not come of its own accord as influence of the moving forces on sense, but must be made' (*OP* 22:321). Indirect appearances, understood as objects of experience, must be made by the subject from direct appearances. The indirect appearance is ultimately an appearance of the direct appearance. If direct appearances are of the dynamic ether or what I have called "Substance," indirect appearances are of objects of experience or what I have called "substances," i.e., what appears in empirical intuition after the subject has organized direct appearances. As Kant says in *Convolut* 10:

> The perception of outer objects is nothing other than the act of the subject through which it affects itself and perceptions are nothing other than moving forces combined together with consciousness.
>
> (*OP* 22:392.3–6)[6]

In this quote, I believe that Kant is using "perception" in two ways, both of which are familiar from *CPR*. In the Transcendental Dialectic, Kant distinguishes between two different senses of "perception." The first is what I will call "subjective perception", i.e., 'a perception that refers to the subject as a modification of its state is sensation' (*CPR* A320/B376). Although subjective perceptions need not rely on external affection (e.g., pleasure and pain under certain circumstances), subjective perceptions can be the result of external affection as long as nothing objective is represented through the sensation.[7] In contrast, 'an objective perception is a cognition' (*CPR* A320/B376) which requires the unification of singular intuitions and general concepts.[8] Although the ether affects the subject through its moving forces, these are only subjective perceptions or sensations that do not represent anything objective.[9]

As I argued above (§1.2), we do not have empirical intuitions of the ether or Substance since empirical intuitions are singular representations. As I will argue below (§4.2), we cannot subsume the ether under the schematized categories. The ether is not the right kind of thing for objective perception but this does not preclude us from having subjective perceptions of the ether. In fact, Kant argues in *Convoluts* 10–11 that subjective perceptions of the ether or Substance are a necessary (though not sufficient) condition for the objective perceptions of substances. The subject transforms these subjective perceptions of the ether (direct appearances) into the objective perceptions of substances (indirect appearances) by applying the categories (general concepts) to the former. Going back to the penultimate block quote above (*OP* 22:340–341), we can understand the categories as the *a priori* element of indirect appearances. Kant also claims, in this quote, that the subject can extract from intuition only what the subject has inserted into appearance. What are inserted, I would suggest, are the categories, which bring unity to direct appearances and generate indirect appearances. This is quite similar to Kant's claim in the A-edition Transcendental Deduction that we would not be able to find unity in nature if we did not already put it there ourselves. This is an '*a priori* certain unity' (*CPR* A125) since it is the result of subsuming appearances under *a priori* categories.[10] Subsuming subjective perceptions (direct appearances) under the categories generates objective perceptions (indirect appearances). As Kant says, 'the transition to physics consists, first, in transforming what is subjective in perception into what is objective in the appearance of the object of the senses' (*OP* 22:458).

The objective perception of substances requires that Substance *affects* the subject through its moving forces (direct appearance) and that the subject *affect* itself by combining direct appearances into spatially discrete substances (indirect appearance).[11] Once these substances are generated, however, they can stand in locomotive affective relations (familiar from *MFNS*) with cognitive subjects which are substances in their own right insofar as they are embodied. I will return to these ideas again in the next chapter since I believe the distinction between direct and indirect appearances is vital to understanding how Kant can overcome the problem of affection.

Even though we do not experience space itself as an object, that the omnipresent forces of the ether affect us is absolutely necessary for our experience of particular objects in a unified spatial framework. Put in terms of the Analogies, even though the ether is not a spatiotemporal object of experience, it serves as the substrate for the spatiotemporal determination of these objects. As Kant says in *Convolut* 11:

> Experience has at its basis (1) *perception*—which always requires moving forces affecting the subject (be they outer or inner) (2) [that] the perceived be elevated to *experience*. For which an inner principle of the subject is required, to think the perceived object in its thoroughgoing determination. For whatever we have experience of there is required a formal principle of thoroughgoing determination.
>
> (*OP* 22:499)

Kant's use of "thoroughgoing determination" in this passage should remind one of his discussion in the Transcendental Ideal of *CPR* (see §1.3) and the *Übergang* section (see §3.3). Just as the totality of empirical reality (or Substance) is necessary for the thoroughgoing determination of objects of experience (or substances) in the Transcendental Ideal without itself being subject to the principle of thoroughgoing determination, so too is the ether necessary for the thoroughgoing determination of objects of experience in Kant's post-Critical transition project without itself being subject to the principle. The ether cannot itself be an object of experience for the reasons mentioned above, but it does make space perceptible and is a necessary (though not sufficient) condition for the thoroughgoing determination of objects of experience.[12] The other necessary (and jointly sufficient condition) for this thoroughgoing determination is the application of the ether concept by the subject, i.e., the "formal principle" Kant mentions in the above quote. We will return to the complementary roles of the material and formal conditions of experience in generating objects of experience (i.e., substances) below and in the final chapter.

In addition to the role the ether plays in the thoroughgoing determination of objects of experience (Transcendental Dialectic) and in providing the substrate for the spatiotemporal unity of the subject's experience (Transcendental Analytic), it also has an important role to play in ensuring the empirical reality of space and time (Transcendental Aesthetic). Although the formal transcendental character of space and time remain unchanged in *OP*, the conditions of their empirical reality are modified by the conclusion of the Ether Deduction. As Kant says in *Übergang* 2:

> The proposition: "There are physical bodies" presupposes the proposition: "There is matter whose moving forces and motion precedes the generation of a body in time." . . . Thus there must exist a matter which, {as internal, penetrates all bodies (as onus), and, at the same time, moves them continually (as *potentia*). . . . The ground for this assertion is: Intuitions in space and time are mere forms, and, lacking something which renders them knowable for the senses, furnish no real objects whatsoever to make possible an existence in general (and, above all, that of magnitude). Consequently space and time would be left completely empty for experience.
>
> (*OP* 21:216–217)

Later, in *Convolut* 10, Kant claims that the ether is 'presented to sense as (empirically determined) space not as derived from or out of experience but rather thought for the possibility of experience' (*OP* 22:316.11–15).[13] This idea is not completely new. In fact, Kant talks about matter making empirical space sensible in *MFNS*.[14] It is no surprise, furthermore, that in the *Elementarsystem,* Kant characterizes the ether as imponderable, incoercible, incohesible, and inexhaustible, since all these are characteristics of space as well.[15]

In these passages from *Übergang* 2 and *Convolut* 10, I believe Kant is returning to a distinction, central to the Transcendental Aesthetic, between the transcendental ideality and the empirical reality of space and time. In the above quote from *Übergang* 2, Kant holds that the ether is a necessary condition for the existence of any empirical object in space and time and as such is a necessary condition for the empirical reality of these objects in space and time. Even so, this does not change the transcendental ideality of space as an *a priori* form of intuition.[16]

It is the unification of space as an *a priori* intuition and the ether as the empirical determination of space that allows objects to be related to one another in a single spatial framework. Although Kant believes that the ether, notwithstanding its omnipresence, affects the subject in *OP*, it is important to note that the ether also post-Critically underwrites his theory of affection in *MFNS*. Without the ether, the embodied subject could not be externally related to locomotive bodies (substances) in a three-dimensional framework as Kant's theory of affection in *MFNS* requires (see §1.2). Ultimately, the ether as material condition and space as formal condition must be *combined* if there is to be a unity of experience. Kant suggests that the unity of space through the ether constitutes a foundation for the unity of experience:

> The unity of space is a foundation of the unity of all outer experience when the discussion is of matter generally and it is actually not experiences rather only the outer perceptions which all belong to one possible experience.
>
> (*OP* 22:287.1–4)[17]

One might wonder after all this talk of space, how the ether and time relate to one another. Kant does not say much about time specifically in *Convoluts* 10–11. We have already noted that Kant considers Substance (in the Analogies) or the ether (in the *Übergang* section) to be both omnipresent and sempiternal. Consequently, it has a role in insuring not only the empirical unity of spatial experience through its omnipresence, which rules out the experience of empty space, but also has a role in insuring the empirical unity of temporal experience through its sempiternality, which rules out the experience of empty time. At this point, Kant probably considered the connection unproblematic, which leads him simply to generalize his discussion of the one form of intuition to the other:

> All the moving forces of matter constitute through the unity and collective relations in space and time a whole under principles of the motion of matter and also their affection of sense—the coming together of elementary concepts into a system of physics.
>
> (*OP* 22:293.16–20)[18]

Assuming that Kant's discussion of space extends to time, one ought to say that the *material* condition for the empirical reality of time is the same as that of space. The addition of this material condition marks a significant revision of Kant's account of space and time in the Transcendental Aesthetic of *CPR*, but is thoroughly consistent with what Kant has to say in the Analogies of the Transcendental Analytic. When it comes to the latter, what is most important to note is that the moving forces of matter as well as the forms of intuition are *necessary* conditions for the empirical unity of spatiotemporal experience. As the quote above makes clear, however, Kant hopes to achieve more than this. He also wants to establish a system of physics by bringing together the transcendental *material* condition for experience with the transcendental *formal* conditions for experience. This is an important part of Kant's attempt to effect a transition from the metaphysical foundations of natural science to physics which, if successful, should bridge the gap in Kant's Critical philosophy.

4.2 THE CATEGORIES IN THE TRANSCENDENTAL ANALYTIC

The addition of a transcendental material condition of experience in the Ether Deduction requires dramatic revisions to the way that the transcendental formal conditions of experience function empirically. Although the forms of intuition have already been dealt with, Kant goes on to discuss how the other transcendental formal conditions of experience must be revised in their empirical function. One goal is to establish the foundations of a system of physics itself *a priori*. Kant offers many definitions of "physics" in *Convoluts* 10–11. Some of these definitions have to do specifically with the moving forces of matter, e.g., 'Physics is the doctrinal system of the moving forces of matter, insofar as it can be exhibited (*exhiberi*) in experience' (*OP* 22:511).[19] Other definitions suggest that physics relates to a subject's cognition more generally, e.g., 'Physics is empirical cognition in a system' (*OP* 22:339.13).[20] Although the former definition of physics might seem far more germane to natural science than the latter, for Kant the two are not really that different at all:

> So as experience is an absolute unity (only to think in *singulari*) so is matter in general also an absolute unity and one can just as little speak of **materials** as of experiences simply because this concept contains merely the *form* of experience and the affection of sense through moving forces in general and this concept is therefore identical with the concept of experience.
>
> (*OP* 22:471.3–8)[21]

Here Kant seems to be equating the concept of the ether with the concept of experience. This might seem strange, but it is important to remember the source of Kant's concept of the ether in the Ether Deduction. Kant discovered the concept of the ether by analyzing the concept of the unity of experience. In *Convoluts* 10–11, furthermore, the *a priori* concept of the ether takes on the harmonizing function that the concept of the unity of experience possessed in the Ether Deduction. Much as experience is a systematic unity (*one* experience), so too is the concept of the ether or Substance the concept of a systematic unity as articulated in the *Elementarsystem* and elsewhere in *OP* (*one* matter). This systematic unity of the moving forces of matter is, in turn, a necessary condition for the systematic unity of experience. The transcendental conditions of experience, however, as they stand at the end of the Ether Deduction, are insufficient for establishing the systematic unity of experience. As mentioned at the end of the last chapter, while the formal transcendental conditions of experience and the material transcendental condition of experience are individually necessary for the unity of experience, Kant must establish their joint sufficiency by showing that there is no *in principle* discontinuity between them (i.e., between the world as thought *a priori* through these formal conditions and the world as given *a posteriori* through this material condition).[22] Without this, Kant cannot effect a transition from the metaphysical foundations of natural science to physics and so cannot fill the gap that lies at the very heart of his Critical philosophy. Kant's broad, post-Critical conception of physics requires establishing the correspondence between the *one* experience (subjective) and the *one* matter (objective). It is important to note how much broader Kant's conception of "physics" is in *Convoluts* 10–11 from his conception in *MFNS*.[23] Whereas the latter limits physics to a doctrine of body, the former expands physics into a post-Critical doctrine of experience, i.e., 'physics as the universal doctrine of experience' (*OP* 22:488). Consequently, the metaphysical foundations of natural science are no longer simply providing a 'metaphysical doctrine of body' (*MFNS* 4:473), but instead are providing a metaphysical foundation for experience itself.

One might argue that since Kant is changing his conception of "physics" from the conception he adopted in *MFNS*, the *OP* does constitute a correction of *MFNS*. If this is true, then wouldn't the gap in Kant's Critical philosophy really lie in *MFNS* contrary to what I claimed in §0.4? The problem with this interpretation is that it assumes that any correction to *MFNS* must be because of a problem *internal* to *MFNS*. My claim, however, is that this correction to *MFNS* results from Kant's discovery of a problem with the Critical philosophy *external* to *MFNS*. As I have argued, the gap in the Critical philosophy lies within the transcendental part of the metaphysics of nature, not within the special metaphysical natural science that Kant develops in *MFNS*. Since the latter presupposes the former, it should not be surprising that a change to the former has ramifications for the latter. Crucially, however, it is not a gap within *MFNS* that has ramifications for the completeness of Kant's view in *CPR*, but rather that a gap within *CPR* that has ramifications for Kant's view in *MFNS*.[24]

Just as with "physics," Kant also formulates many different definitions of "transition" in *Convoluts* 10–11. At points, Kant suggests the transition should be to physics in the sense of the moving forces of matter, e.g., 'The transition from the metaphysical foundations of natural science to physics proceeds through the *actus* of combining the moving forces of matter in accordance with principles *a priori* into a system of natural research in general' (*OP* 22:285:1–3).[25] At other points, however, Kant thinks the transition should be to physics understood more generally in terms of the unity of the subject's representations. For example, Kant says:

> The transition from the metaphysical foundations of natural science to physics is therefore not an aggregation of empirical representations with consciousness, rather the concept of the synthetic unity of these representations for the possibility of experience which at any time is thought as a system of empirical representations (not an empirical system for that would be a contradiction).
>
> (*OP* 22:359.20–25)[26]

Just as Kant did not see his different definitions of "physics" as being different in kind, so too he does not see these different definitions of "transition" as being different in kind. The systematic unity of the subject's representations (*one* experience) is inextricably linked to the systematic unity of the moving forces of matter (*one* matter). They are two sides of the same coin which must correspond to one another. The transition is the means by which "physics" in the sense of coordinated cognitive (*one* experience) and material (*one* matter) unity is established. Unlike early leaves of *OP*, the transition is no longer simply the application of *MFNS* to empirical physics.

Throughout *Convoluts* 10–11, Kant makes clear that the ether will play an important formal and material role in effecting the transition and thereby establishing "physics" in the above sense. Given the addition of the ether concept as a formal condition of experience, it is important to examine what changes Kant makes to the Transcendental Analytic in order to accommodate this concept. As mentioned in Chapter Three (§3.3), it is important to note that, in the *Übergang* section, the concept of the ether must be *a priori* since the proof of the ether is itself *a priori*. It is no easy task, however, incorporating the *a priori* concept of the transcendental material condition of experience into the framework of *CPR*. This concept cannot be a category. Categories are concepts of an object in general which take appearances in empirical intuition and determine them into objects of experience.[27] The ether *cannot* be an object of experience, however, on pain of undermining the *a priori* proof strategy of the Ether Deduction. As we saw in Chapter One (§1.3), furthermore, Kant requires two *a priori* concepts of substance if he is to avoid the dilemma that faces his theory of substance in the Analogies. Although the category of substance can be understood as applying to substances, Kant requires an *a priori* concept of Substance (or the ether) distinct from the category of substance. Likewise, the concept of the ether cannot be an empirical concept. Since the ether is not an object of experience, it is unclear what empirical marks could constitute the concept. This problem is compounded by Kant's thoroughly negative characterization of the ether in the *Elementarsystem* with respect to the properties that substances (which are objects of experience) might possess. Likewise, one must also remember that the ether is necessary for the thoroughgoing determination of objects of experience without itself being subject to the principle of thoroughgoing determination. Even so, Kant argues that the transition from the metaphysical foundations of natural science to physics requires not only the actuality of the ether (something we know from the Ether Deduction), but also the 'recognition' of this material which requires *a priori* concepts. As he says in *Convolut* 11:

> This radical world-material is not problematic and merely assertoric, but apodictically certain. Its existence belongs to the transition from the metaphysical foundations of natural science to physics; and its recognition (according to *a priori* concepts of objects in appearance in general (regarded not *sparsim* but *coniunctim*)) makes physics initially possible, according to the principle of the possibility of experience, which is itself only single, and, objectively forms a system.
>
> (*OP* 22:475–476)

Here Kant seems to be tying the recognition of the ether to the application of the categories themselves, i.e., '*a priori* concepts of objects in appearance in general.' Since the concept of the ether is not a category, however, how do we make sense of what Kant is saying? At this point, it is important to recall that the concept of the ether that Kant deploys in *Convoluts* 10–11 (as well as throughout most of the transition project) is that of an elementary system of the moving forces of matter which he aims to show has an *a priori* constitutive function in experience. My contention is that this concept is equivalent to Kant's concept of the *ether* which is a *complex* concept, one that comprehends both the functions Kant attributes to the ether (§2.0) as well as its (positive and negative) categorical characterization (§2.2). One can see Kant's conception of the ether becoming more detailed as he moves out of the Critical and into the post-Critical period of his philosophy. Whereas in *CPR*, his concept of Substance was of something sempiternal and omnipresent that insures the empirical unity of the subject's spatiotemporal experience. In *OP*, the ether becomes a plenum of dynamic (attractive and repulsive) forces that has certain *functions*. It is the source of affection, body formation, and mechanical motion, while also unifying space and time. It is *positively* characterized as universally distributed in space (quantity), all-penetrating (quality), internally moving (relation), and physically necessary (modality). At the same time, it is *negatively* characterized in terms of all of the categorical properties that substances possess. It is imponderable (quantity), incoercible (quality), incohesible (relation), and inexhaustible (modality). It is this mature *Elementarsystem* of the ether or Substance that serves as Kant's *a priori* concept for transition in *Convoluts* 10–11. Although some of these functions and properties are only implicit until the *Übergang* section, it stands to reason that Kant is referring to all of these functions and properties when uses the term "*Elementarsystem*" later in *Convoluts* 10–11.

In *Convolut* 10, Kant claims that the categories serve as the *a priori* basis for the *Elementarsystem*: 'The elementary concepts of the moving forces of matter . . . stand under the categories' (*OP* 22:342). Kant also emphasizes the important role that this *Elementarsystem* has to play in unifying experience. As he says in *Convolut* 11, 'The *Elementarsystem*—a formal principle of the unification of the moving forces to a unity of experience' (*OP* 22:339.15–16).[28] Kant believes that this concept precedes experience *a priori* in the understanding and has an important *formal* role to play in the transition project. In a loose leaf of *Convoluts* 10–11 discovered after the publication of *Kants gesammelte Schriften,* Kant mentions the concept of an elementary system of the moving forces of matter in just this context:

> The tendency of the metaphysical foundations of natural science to physics as a doctrinal system of the principles of the same contains *a priori* in itself the concept of an *Elementarsystem* of the moving forces of matter as its form.[29]

In what follows, and using the definitions of "transition" and "physics" above, I will argue that the application of this concept to objective perceptions serves to complete the transition from the metaphysical foundations of natural science to physics, which will also fill the gap in Kant's Critical philosophy. The ether is a necessary though not a sufficient condition for objects of experience. As we saw above (§4.1), the moving forces of matter generate subjective perceptions (sensations) that the subject transforms into objective perceptions of substances (objects of experience). The subject organizes direct appearances of the ether in accordance with the categories to transform them into indirect appearances (objects of experience) in empirical intuition. As I will argue in the next chapter, Kant continues to view these objects of experience or substances as standing in mechanical (locomotive) affective relations with embodied cognitive subjects. Crucially, what we objectively perceive in empirical intuition is not the ether or Substance itself, but rather relatively persistent and spatially discrete objects of experience (substances) that the subject has constructed in empirical intuition and which stand in affective relations to the subject. The categories have proper application to these substances since the categories constitute them. In the Transcendental Deduction, Kant defines an "object" as 'that in the concept of which the manifold of a given intuition is united' (*CPR* B137). The manifold of a given empirical intuition is united into an object of experience through the categories. Without the categories, there would be no objects of experience or substances according to Kant. Although Kant explains in the Transcendental Analytic how the categories (in their schematized form) are applied to empirical intuition for the possibility of objects of experience, how does the concept of the ether serve as a condition for the possibility of experience?

Kant's answer in *Convoluts* 10–11 is to say that the objective application of this concept for experience is effected through the Principles of Pure

Understanding. Unlike in *CPR,* where the Principles are objectively valid on their own, in *Convolut* 10 Kant sees the Principles as effecting a transition from subjectivity to objectivity. Objectivity itself, however, is achieved only when the concept of the ether has been applied *through* the Principles resulting in the unity of experience. This reveals the *formal* role that the concept of the ether has as the *a priori* concept for transition. As Kant says in *Convolut* 10:

> Transition to the concept: 1) Axioms of Intuition. 2) From the intuition to perception, perception to experience Analogies. 3) *subjective*—4) Transition to the unity of experience in one system of forces objective.
> (*OP* 22:289.20–23)[30]

Within the context of *CPR,* the Principles are rules for the objective application of the categories, but in *OP* they also serve as rules for the objective application of the concept of the ether.[31] As mentioned above, the concept of the ether is not a category or an empirical concept. It is not a mere problematic concept or transcendental idea either since the object of this concept (ether) is meant to be *actual* given the conclusion of the Ether Deduction. Hence, this concept has a totally unique position within Kant's post-Critical system.[32] Even so, Kant leaves room for such a concept within the Critical philosophy. Whereas an empirical concept is 'borrowed from [experience],' a pure concept is 'one on which, as *a priori* condition, experience in general (its form) rests' (*CPR* A220/B267). As an *a priori* conceptual condition for the unity of experience, the concept of the ether meets Kant's conditions for a pure concept. Although it is not a category, Kant allows for other concepts (e.g., mathematical concepts) to be pure concepts within the context of *CPR.*[33]

Whereas the concept of the ether is characterized through the categories in the *Elementarsystem,* this concept is applied by the categories, in their role as Principles, in *Convoluts* 10–11. With regard to the Principles, in *Convoluts* 10–11, Kant discusses the Axioms of Intuition (quantity), Anticipations of Perception (quality), and the Analogies of Experience (relation).[34]

Although Kant does not explicitly mention the Postulates of Empirical Thought (modality) in the above passage, and does not discuss them at any length in *Convoluts* 10–11, the way in which the concept of the ether is applied for experience nevertheless follows the lessons of the Postulates. As mentioned at the end of the last chapter, Kant's discussion of the ether is consistent with what he has to say in the Postulates of Empirical Thought and the latter can help him avoid the charge of transcendental realism in the Ether Deduction. Even though Kant spends little time discussing the Postulates in *Convoluts* 10–11, it might be useful to examine, on his behalf, how *Convoluts* 10–11 could make use of the Postulates.

In *CPR,* Kant discusses the Second Postulate's application to the case of "magnetic matter" which though not perceived is nevertheless actual in accordance with natural law insofar as its effects (the movement of iron filings) are perceived. In this way, Kant says that 'one can also cognize the existence of the thing prior to the perception of it' (*CPR* A225/B273).[35] Kant's Ether Deduction in the *Übergang* section and his transition project in *Convoluts* 10–11 proceed in much the same way. In addition to arguing that the ether is actual, however, Kant claims in the *Übergang* section and *Convoluts* 10–11 that it is also necessary. When going through the negative properties of the ether in *Convolut* 11 (i.e., imponderable, incoercible, incohesible, inexhaustible), he says that 'here modality contains the category of necessity' (*OP* 22:459). Likewise, in *Redactio* 1–3, written contemporaneously with the early sections of *Convoluts* 10–11, Kant again notes the necessity of the ether. In addition to his negative characterization of the ether as imponderable (quantity), incoercible (quality), incohesible (relation), and inexhaustible (modality), under the category of modality, Kant also emphasizes the ether's '*necessity in the appearance*' (*OP* 22:583.16).[36] It is important to recall that the ether is a transcendental *material* condition of experience. Given what Kant says in the Third Postulate, the ether must be *materially* necessary (for existence), but not *logically* necessary (for the connection of concepts).[37]

Returning to the above passage from *Convolut* 10 which mentions the other Principles, Kant's crucial point is that a *subjective* progression from intuition (Axioms) to perception (Anticipations) to experience (Analogies) is required for the *objective* application of the concept of the ether *for* experience. When it comes to the Axioms and Anticipations, Kant believes these Principles correspond to the extensive (Axioms) and intensive (Anticipations) magnitudes of the moving forces of matter.[38] In *Convoluts* 10–11, Kant continues to maintain that the subject is affected by the moving forces of matter, though he believes that the subject anticipates these moving forces *a priori*.[39] The moving forces of matter generate extensive magnitudes in space and intensive magnitudes in time.[40] Consequently, the subject can be certain, given the Axioms and Anticipations, that every empirical object, which results from the application of the categories of quantity and quality, will be composed of parts (extensive magnitude) and will have a degree of influence on sense (intensive magnitude).

Whereas the Axioms and Anticipations are mathematical principles that pertain merely to intuition and as such are intuitively certain and unconditionally necessary, the Analogies and Postulates are dynamical principles that are necessarily conditional upon some existent given in intuition.[41] In an important way, the Analogies and Postulates presuppose what is already given through the Axioms and Anticipations. When it comes to the Analogies, they are responsible for uniting the *perceptions* gained through empirical *intuition* into the *experience* of objects.

As mentioned above, the radical difference between Kant's use of the Principles in *CPR* and his use of the Principles here in *Convoluts* 10–11 is that the Principles are themselves objectively valid of experience in the former, whereas they are only part of a subjective transition in the latter. The subject organizes subjective perceptions of the ether into objective perceptions of substances using the Principles. These objective perceptions, however, are not united into a *single* experience until application of the *a priori* concept of the ether.

Although objective perceptions are multitudinous and multifarious, there is only *one* unity of experience. Furthermore, the singularity of experience cannot be achieved without the concept of the ether, i.e., the concept of *one* matter. Kant believes that without the absolute unity of consciousness that this concept makes possible, there would be only a fragmentary distributive collection of objective perceptions which is itself insufficient for experience. As Kant says in *Convolut* 10:

> Experience is absolute unity of the consciousness of the reality of a sensible object and there is only one experience. When experiences are spoken of so one understands thereby only perceptions (empirical representations so far as they are aggregated with one another) which lack much to be raised to the validity of one experience and to be established as belonging to physics because this should be a system, which can expect its truth only from the harmonization [*Zusammenstimmung*] of all perceptions united in one whole of the same which cannot occur fragmentarily.
>
> (*OP* 22:280.21–28)[42]

Given Kant's definition of "physics" above, one should understand it simply as the correspondence between the absolute unity of consciousness (*one* experience) and the systematic unity of the moving forces of matter (*one* matter). What accounts, however, for the correspondence between the subjective (consciousness of the world) and the objective (world conditioned by the ether)? Establishing this correspondence is a task that Kant saw for himself already in *Übergang* 12:

> Nevertheless, the idea of this [system] is unavoidably given *subjectively*—as a necessary problem, namely, that of the connection of perceptions as effects of the moving forces upon the subject in a *single* experience. What, however, belongs to experience (which can only be single) as its ground of determination, is likewise *objectively* given—that is, *actual*. So there *exists*, as an absolute whole, a matter with those attributes, as the *basis* of its moving forces, insofar as they are moving.
>
> (*OP* 21:601)[43]

Kant's answer in *Convoluts* 10–11 is that whereas the moving forces of matter produce subjective perceptions in the subject, this is insufficient for a single experience. The subject must *insert* its *a priori* concepts into these subjective perceptions in order to generate the objective perceptions of substances in a single spatiotemporal framework, i.e., a single experience. This comes out clearly, I believe, in a passage which I earlier quoted only in part (see §3.4):

> The first principle of representation of the moving forces of matter [is] to regard them not as things in themselves but as phenomena, according to the relation which they have to the subject—as they affect our sense, or as we affect our sense ourselves. [It involves] **inserting** the formal element of sensible representation into the subject in order to progress from the Axioms of Intuition, the Anticipations of Perception, etc. to experience—that is, for experience as a system, not as derived from experience. . . . this is the formal unity of experience as a system of perceptions. The material unity of experience is the idea of a whole of moving forces as the absolute (unconditioned) unity of the world-system, in which the moving forces contain and initiate nothing outside this *complexus*.
>
> (*OP* 22:300)[44]

It is the insertion of the categories in their role as Principles (subjective perceptions to objective perceptions) that makes possible the application of the *a priori* concept of the ether (objective perceptions to a single experience). The result is a correspondence between the absolute unity of consciousness conditioned formally by the *a priori* concept of the ether (*one* experience) and the world consciousness represents conditioned materially by the ether (*one* matter). Consciousness is harmonized with the moving forces of matter since the ultimate conceptual condition of consciousness just is the concept of these moving forces. Ultimately, it is the application of the *a priori* concept of the ether *for* experience that completes the transition by establishing his post-Critical conception of physics.

What remains to be explained is how exactly this insertion occurs. In the next section, I will examine the subject's activity of *insertion* which I attribute to apperception. I contend that post-Critical apperception serves not only to explain the subject's activity in *Convoluts* 10–11, but also to obviate those problems which attend Kant's conception of apperception in the Critical period.

4.3 MIND, WORLD, AND APPERCEPTION

Kant's desire to establish the correspondence between mind and world is not new to *OP*. Demonstrating the harmony between the systematic unity of cognition and the systematic unity of nature *a priori* (his post-Critical conception of physics) is of utmost importance for the Critical philosophy though, as I will argue, the Critical philosophy ultimately fails to establish such a harmony. In the A-edition of the Transcendental Analytic, Kant claims that the order and regularity of appearances is a result of the subject actively *inserting* this order and regularity into appearances *a priori*. The product is the unity of nature:

> Thus we ourselves bring into the appearances that order and regularity in them that we call nature, and moreover we would not be able to find it there if we, or the nature of our mind, had not originally **inserted** [*hineingelegt*] it there. For this unity of nature should be a necessary, i.e., *a priori* certain unity of the connection of appearances. But how should we be able to establish a synthetic unity *a priori* if subjective grounds of such a unity were not contained *a priori* among the original sources of cognition in our mind, and if these subjective conditions were not at the same time objectively valid, being the grounds of the possibility of cognizing any object in experience at all?
>
> (*CPR* A125)[45]

Kant makes related claims throughout *CPR*. In the Preface to the B-edition of *CPR*, Kant says that 'we cognize of things *a priori* only what we ourselves have put into them' (*CPR* Bxviii). One example of this, which I discussed earlier, is the categories. The categories are the concept of an object in general and allow us to think of objects *a priori*.[46] This thinking requires a certain activity on the part of the subject which Kant attributes to transcendental apperception. When describing the role of transcendental apperception as transcendental ground in the A-edition Transcendental Deduction, Kant says:

> Every necessity has a transcendental condition as its ground. A transcendental ground must therefore be found for the unity of the consciousness in the synthesis of the manifold of all our intuitions, hence also of the concepts of objects in general, consequently also of all objects of experience, without which it would be impossible to think of any object for our intuitions; for the latter is nothing more than the something for which the concept expresses such a necessity of synthesis.
>
> (*CPR* A106)

While serving as ground for the unity of consciousness in *general*, transcendental apperception simultaneously produces the unity of nature in *general*.[47] Even so, it cannot guarantee the *systematic* unity of cognitions (the unity of all possible cognitions) and so produce the *systematic* unity of nature (the unity of all possible manifolds of appearance).

Kant discusses the systematic unity of cognitions in the appendix to the Transcendental Dialectic. He says there that the systematic unity of the understanding's cognitions can only be a *projected* unity, i.e., a mere idea of reason which cannot be produced by the understanding itself. Reason must attempt, nevertheless, systematically to unite its findings with understanding's cognition of nature.[48] Since transcendental apperception cannot produce a systematic unity of the understanding's cognitions, it also cannot produce the systematic unity of nature. Reason can only *presuppose* such a unity as objectively valid and necessary. This presupposition is itself necessary, however, since without it there would be no coherent use of the understanding:

> For the law of reason to seek unity is necessary, since without it we would have no reason, and without that, no coherent use of the understanding, and, lacking that, no sufficient mark of empirical truth; thus in regard to the latter we simply have to presuppose the systematic unity of nature as objectively valid and necessary.
>
> (*CPR* A651/B679)

Kant's claim here should give the reader pause. If there is 'no sufficient mark of empirical truth,' i.e., no guarantee of the 'agreement of cognition with its object' (*CPR* A58/B82), Kant's whole theory of cognition in *CPR* might be undermined.[49] As an illustration, consider empirical propositions concerning events. In order for these empirical propositions to have truth-conditions, Kant would hold that the events these propositions describe must be incorporated into a unified system. Take any empirical proposition that describes an event in the natural world. According to the Second Analogy, in order for something to be an event, it must assume another event from which it follows in accordance with natural law. The same is true of the latter event, and so on. Consequently, in order for the empirical proposition to be true, it must assume that the event it describes is part of a complete network of events connected with one another in accordance with natural law. In other words, the truth of the individual empirical proposition assumes the systematic unity of nature. Although this systematic unity cannot be demonstrated constitutively but can only be assumed regulatively, it must be assumed nonetheless.[50]

Are we justified, however, in assuming this systematic unity? The general worry can be expressed through two closely related problems. (1) There is a "top-down" problem for Kant's theory of experience. There may be a gap between the categories and appearances insofar as there could be sets of appearances that cannot be fully determined by means of the categories into cognitions of objects.[51] (2) There is a "bottom-up" problem for Kant's theory of experience. There could be rogue objects which are capable of affecting the subject, producing appearances in sensibility, though these appearances are not fully determinable by means of the categories into cognitions.[52] If there are manifolds of appearance that cannot be determined by means of the categories, this would undermine the systematic unity of nature and would preclude us from presupposing this unity as objectively valid and necessary. Without the systematic unity of nature, however, there is no sufficient mark of empirical truth, i.e., no guarantee that cognition agrees with its object. One might be tempted to dismiss these undeterminable manifolds as mere *hyperbolic* possibilities. From the perspective of human understanding, however, Kant is seriously concerned that there could be undeterminable manifolds of appearances in *CJ:*

> But it does not follow from this [that experience is a system in accordance with transcendental laws] that nature even in accordance with **empirical** laws is a system that **can be grasped** by the human faculty of cognition, and that the thoroughgoing systematic interconnection of its appearances in one experience, hence the latter itself as a system, is possible for human beings. For the multiplicity and diversity of empirical laws could be so great that it might be possible for us to connect perceptions to some extent in accordance with particular laws discovered on various occasions into one experience, but never to bring these empirical laws themselves to the unity of kinship under a common principle, if namely, as is quite possible in itself (at least as far as the understanding can make out *a priori*), the multiplicity and diversity of these laws along with the natural forms corresponding to them being infinitely great, were to present to us a raw chaotic aggregate and not the least trace of a system, even though we must presuppose such a system in accordance with transcendental laws.
>
> (*CJ* 20:209)

Although Kant admits that, from the perspective of our understanding, there could be undeterminable manifolds in *CJ,* he also understands that establishing the systematic unity of nature is essential for his theory of cognition. Kant's solution is to deploy the reflective power of judgment as the faculty that stipulates *a priori* a transcendental principle for nature as a condition for bringing about the unity of nature's empirical laws.[53] Even so, the unity is merely presupposed here, just as the systematic unity of nature was presupposed in the appendix to the Transcendental Dialectic.[54]

Kant's project changes in *Convoluts* 10–11 insofar as he believes that he need not merely *presuppose* but can actually *establish* a correspondence or harmony between the subject's cognitions and the systematic unity of nature *a priori.* The systematic unity of nature is itself deduced *a priori* in the Ether Deduction. The systematic unity of the subject's cognitions via the absolute unity of consciousness is established in *Convoluts* 10–11, as is the harmonization of this absolute unity of consciousness with the systematic unity of nature. Kant's project, if successful, would dispel the top-down/bottom up problems since there would be no *in principle* discontinuity between the systematic unity of nature and the subject's cognition of this systematic unity. Dispelling this problem would safeguard Kant's theory of cognition in the Transcendental Analytic by establishing the sufficient mark of empirical truth which Kant seems to lack in the appendix to the Transcendental Dialectic. There is a serious *systematic* problem for the Critical philosophy, however, if Kant is not successful in *Convoluts* 10–11.

Just as apperception plays a large role in every phase of the Critical project, so must apperception continue to serve an important role post-Critically insofar as it is required to overcome the above problem. Much of what Kant has to say about apperception in *CPR,* he repeats again in *Convoluts* 10–11. He holds that it is a combinatorial faculty that systematizes the subject's representations.[55] Like the A-edition of the Transcendental Analytic, Kant believes, in *OP*, that 'we can research in nature nothing other than what we **insert**' (*OP* 22:286.10–11).[56]

Some philosophers take quotes like the last one and argue that Kant is well down the road toward a kind of absolute idealism.[57] It might seem that, through apperception, the subject constitutes space and time as well as its manifold. If this were the case, then the subject would posit not only itself as an object in this manifold, but also all of the objects in it. The problem with this view is that it completely ignores the subject's *receptivity* and the role the moving forces of matter have in affecting the subject in sensibility.[58] Although Kant spends a significant amount of time discussing the subject's *activity* through apperception, this should not entail that the subject's *receptivity* has been abandoned. To the contrary, I will argue that both of these components are absolutely essential if Kant is to effect the transition that he envisions. In the next chapter, I will discuss at greater length the role that apperception plays in the constitution of the objects of experience, though I will not ignore the role the moving forces of matter have to play in affecting the subject.

Although it has already been established that apperception is the transcendental ground for the activity of insertion *a priori,* apperception is, within the context of the Critical project, capable of bringing about only a distributive unity of perceptions. Kant claims post-Critically, however, that this distributive unity of perceptions is by itself insufficient to meet the goals of the transition in *Convoluts* 10–11. There apperception must perform two tasks simultaneously: (1) it must insert the Principles into subjective perceptions, generating a distributive unity of objective perceptions, and (2) transform this distributive unity of objective perceptions into the absolute unity of consciousness. This is a much stronger sense of unity than the distributive unity of (objective) perceptions in *CPR*. Although Kant does not discuss apperception by name in *Convoluts* 10–11, its role is exposed through Kant's use of the term "spontaneity":

> Physics is the principle containing the subjective aspect of the perception of moving forces at the same time as the objective aspect of connecting these perceptions to ground experience and the spontaneity of this combination, the form accordingly precedes the receptivity of moving forces *a priori* and this serves as a law which is possible only through the relation to a system of the empirical of cognition, not however an empirical system (*contrad. in adjecto*).
>
> (*OP* 22:297.17–24)[59]

Apperception is the cognitive faculty that spontaneously brings about the absolute unity of objective perceptions in consciousness while at the same time conditioning subjective perceptions themselves *a priori*. In *Convoluts* 10–11, objective perceptions are conditioned ultimately by the *a priori* concept of the transcendental material condition of experience (ether). When the distributive unity of objective perceptions is combined within this concept, an absolute unity of consciousness is generated which is itself harmonized with the systematic unity of the moving forces of matter. Although the moving forces of matter are systematic, our objective perceptions are multifarious, so we must insert something into our objective perceptions to make them systematic.[60] What we insert is the concept of the ether. There is no *in principle* discontinuity between the external world and the subject's conscious representation of it, since the ultimate conceptual condition of consciousness is simply the concept of the ultimate material condition of the external world.

Again, post-Critical apperception, through spontaneity, (1) inserts the Principles into subjective perceptions generating a distributive unity of objective perceptions, and (2) inserts the concept of the ether into the distributive unity of perceptions, generating an absolute unity of consciousness. This latter unity corresponds with the systematic unity of the moving forces of matter (i.e., the ether).[61]

The ether is the material condition for the objects of experience and their spatiotemporal unity.[62] In this respect, the ether is the material condition for the absolute unity of consciousness, whereas the *a priori* conceptual conditions are, of course, the categories (via Principles), as well as the concept of the ether. The absolute unity of consciousness, as the final product of synthesis, is the result of the objective application of this latter concept. Since perceptions are themselves ultimately generated by the ether and formally conditioned by the concept of the ether, Kant can guarantee the harmony between the absolute unity of consciousness (objective application of the concept of the ether) and the systematic unity of the moving forces of matter (ether). The below diagram attempts to summarize Kant's project in *Convoluts* 10–11:

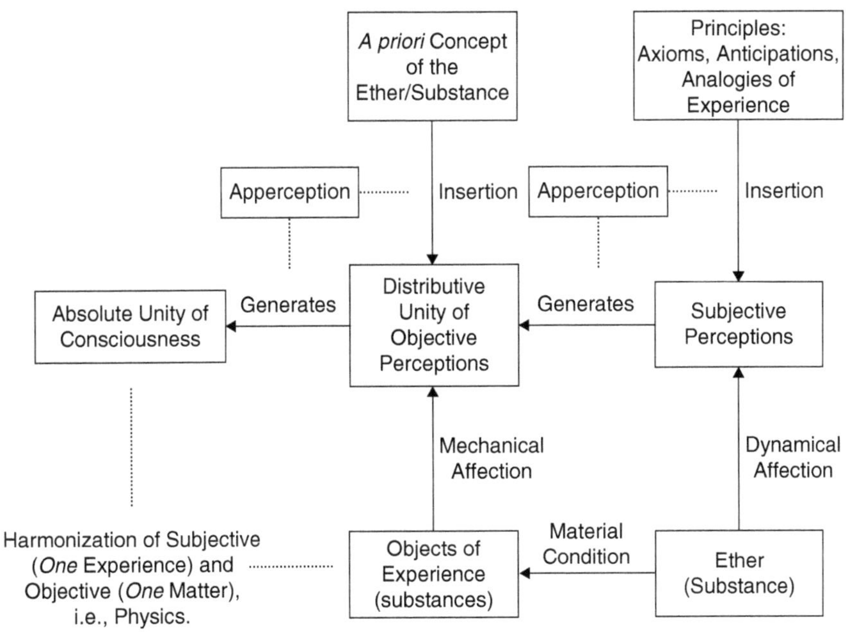

4.4 FILLING THE GAP IN THE ANALOGIES OF EXPERIENCE

At this point we have enough of Kant's post-Critical framework in place to see how he can overcome the dilemma that he faces in the Analogies. As we saw in the first chapter, when Kant talks about substance in the Analogies, he is not always talking about the same thing and one concept of substance is simply insufficient to disambiguate Kant's different and mutually irreducible uses of "substance." On the one hand, Kant must hold that relatively enduring empirical objects are *substances* since they persist through the alteration of their properties. On the other hand, Kant requires one sempiternal and omnipresent *Substance* whose quantum in nature is neither increased nor diminished.

This results in a dilemma for applying the category of substance. If the category of substance applies to Substance, then although the category would capture that single backdrop by reference to which the subject experiences the simultaneity and succession of empirical objects in a common spatiotemporal framework, the subject could not individuate relatively enduring empirical objects and perceive their alterations. In contrast, if the category of substance applies to substances, then although the subject could individuate substances and perceive their alterations, the category would not pick out a single backdrop by reference to which these substances could be experienced as either simultaneous with or successive to one another in a common spatiotemporal framework. Any unequivocal interpretation also fails to make sense of Kant's apparent need for both Substance and substances in the other two Analogies.

Unlike the Analogies, in *OP* Kant clearly delineates between *two* different and mutually irreducible concepts of substance. On the one hand, there is the *a priori* concept of Substance or the ether (where the latter is simply a post-Critical development of the former), and on the other, the category of substance which in its role as a Principle has application to ordinary empirical objects or substances and makes possible the application of the *a priori* concept of Substance or the ether to the dynamic force plenum in conjunction with the other Principles. At several points in *Convoluts* 10–11, Kant even refers to things in space and time as *substances* [*Substanzen*] and to the ether as *Substance* [*Substanz*].[63] With these two *a priori* concepts in hand, Kant is able to overcome the dilemma that faces his theory of substance in the Analogies and so fill the gap in his Critical philosophy.

As we saw above, the *a priori* concept of the ether or Substance is applied *through* the Principles of Pure Understanding. The Principles combine the subjective perceptions of the ether into a distributive unity of objective perceptions of substances. The *a priori* concept of the ether combines this distributive unity of objective perceptions into an absolute unity of consciousness. Put in terms of the dilemma, the category of substance (*qua* Principle) applies to substances insofar as it generates them (in conjunction with the other Principles). This makes possible the application of the *a priori* concept of Substance (ether) which generates a unified spatiotemporal experience of substances. In other words, applying the *a priori* concept of Substance or the ether unifies substances within the spatiotemporal framework that the dynamic force plenum underpins. This concept is applied through an act of apperception which simultaneously combines the distributive unity of objective perceptions into an absolute unity of consciousness. The result is a correspondence between the world (materially conditioned ultimately by the ether or Substance) and the subject's cognition of the world (formally conditioned ultimately by the concept of the ether or Substance). The task of Kant's transition project is to establish this correspondence where the correspondence is itself equivalent to Kant's post-Critical conception of physics.

If Kant's transition project is successful, then he will have explained how the category of substance, functioning in its role as a Principle in the First Analogy, could have application to substances. This in turn makes possible, in conjunction with the other Principles, the application of the *a priori* ether concept, or the concept of Substance for the unity of experience underpinned by the ether. Since substances emerge from Substance or the ether, one can also distinguish between the alterations of these substances (e.g., freshly cut wood weathering), and the arising/perishing of these substances that require alterations of Substance or the ether (e.g., burning wood perishing while smoke and ash arise) by delineating between the category of substance that applies in the former case and the concept of Substance or the ether that applies in the latter.[64] Whereas application of the category of substance allows the subject to individuate substances and perceive their alterations, application of the concept of Substance or the ether insures that all experiences of substances take place in a common spatiotemporal framework. At this point, it is possible to explain, at greater length, how all four of the questions that opened this chapter might be answered:

(1) What is physics? Kant answers that it is 'the aggregate of empirical cognition of the moving forces of matter in experience.' There are two aspects to Kant's post-Critical conception of "physics." The first aspect has to do with how the systematic unity of the moving forces of matter make the subject's experience possible. The second aspect has to do with the systematic unity of the subject's experience conditioned materially by the ether. These two aspects complement one another. Much as there is only *one* experience, so too is there ultimately only *one* matter, and establishing their correspondence is tantamount to establishing "physics" in Kant's post-Critical sense. In what sense can the ether or this "one matter" be considered the ultimate object of physics? Although it is not itself an object of experience, it produces the subjective perceptions necessary for our objective perceptions of physical bodies and makes space thoroughly perceptible. In *Convolut* 11, Kant claims that the ether 'cannot *immediately* be an object of experience' (*OP* 22:480). In this respect, one might consider the

ether to have a similar physical status as magnetic matter does in the Second Postulate (both connected to objective perception without themselves being objectively perceived), though importantly different from an epistemological standpoint insofar as the actuality of the ether is established *a priori*. Even so, the *a priori* concept of the ether is applied to the dynamic force plenum only *indirectly* through the substances we perceive *a posteriori* and to which the categories have application. As I will argue in the next chapter (§5.5), the ether is itself a perpetually undetermined phenomenal object and so can never be experienced in the way that determined phenomenal objects (physical bodies or substances) are experienced.

(2) What is experience? Kant answers that it is 'the unification of perceptions under the principle of their connection (a subjective whole of the aggregate of perceptions), according to concepts, to a doctrinal system.' While the moving forces of the ether produce subjective perceptions, Kant requires the categories (concepts) in their role as Principles to provide a subjective whole of the aggregate of objective perceptions. This distributive unity of consciousness is then conditioned, by an act of apperception, by the concept of the ether. This generates an absolute unity of consciousness (*one* experience) which corresponds to the ether as the material condition of experience (*one* matter).[65]

(3) What is the transition from the metaphysical foundations of natural science to physics? Kant answers that it is 'the doctrinal system of experience in general applied to the natural system.' Just as Kant's conception of "physics" has two aspects, so too does his conception of the "transition." Much as with physics, Kant does not see these aspects as having different ends, but are rather two aspects of the same end. Put simply, the twin purposes of the transition are to establish the system of physics in its two aspects, as a unity of experience (absolute unity of consciousness) which corresponds to a unity of matter (ether). Just as there is only *one* experience, so too is there only *one* matter that produces the perceptions that the subject unifies through the *a priori* concept of this matter into this one experience. When it comes to physics as a system, Kant thinks that it comprehends two kinds of body: organic and inorganic.[66] As mentioned above (§2.3), given the dynamic character of the ether, both inorganic as well as organic bodies can be understood as emerging from it. Consequently, both inorganic as well as organic nature can be unified into one system of nature if Kant's transition project is successful. This marks a departure from *CJ* insofar as the systematic unity of nature is detached from a regulative principle of purposiveness and given constitutive status in *OP*.[67] Although Kant believes organized bodies indicate an 'immaterial principle' for their purposiveness, he describes this immaterial principle as 'a substance, which cannot act otherwise in the distribution of force than as absolute

unity' (*OP* 22:504). He contrasts this substance with an 'aggregate of atoms' which suggests that he may be thinking of the ether (as the systematic unity of the moving forces of matter) as the "substance" at issue. This interpretation is also consistent with Kant's tendency, through much of his career, to contrast his dynamical theory of matter with mechanical theories of matter that rely on atomistic or corpuscularian assumptions.

At this point, it is important to note that Kant never gives up on establishing a *doctrinal* system of physics in the post-Critical period. In other words, Kant aims to fulfill the promise, made in the Preface to *CJ*, to complete his metaphysics of nature. My contention, however, is that this *task* Kant set for himself does not constitute a *gap* in the Critical philosophy. As mentioned in Chapter One (§1.3), Kant makes clear in *Farrago* 1 that he sees the gap as lying in the system from *a priori* principles. This suggests that the gap lies in the *transcendental part* of Kant's metaphysics of nature, i.e., in the Transcendental Analytic. Showing how the pure natural science articulated in *MFNS* can be extended to empirical physics (i.e., completing the doctrinal task) *presupposes* that the transcendental foundations for this pure natural science are themselves complete. As I have argued, the gap in Kant's Critical philosophy lies in the Analogies of Experience of the Transcendental Analytic. By filling this gap in the Principles of Pure Understanding (i.e., in the system from *a priori* principles), Kant is finally in a position to complete his doctrinal task. Although I question Kant's success in completing his doctrinal task (see §2.2), one can see from his post-Critical definitions of "transition" and "physics" that his conception of the "transition from the metaphysical foundations of natural science to physics" includes much more than just the extension of *MFNS* to empirical physics. Unlike early leaves of *OP*, the transition has come to encompass Kant's whole theory of experience.

4) How is the transition from one (metaphysical foundations of natural science) to the other (physics) possible? Kant answers '(i) in respect of the material element of the object, (ii) of the formal element of the subject.' In *OP*, the ether serves both an important transcendental *material* as well as *formal* role. In the Ether Deduction, Kant proves *a priori* the existence of the ether as a transcendental *material* condition for experience. In *Convoluts* 10–11, Kant shows that the *a priori* concept of the ether serves an important conceptual role as a transcendental *formal* condition of experience.

The *a priori* concept of the ether is applied through the Principles (e.g., the Analogies) for the unity of experience. This is made possible by apperception first inserting the Principles into subjective perceptions of the ether, generating a distributive unity of objective perceptions of substances. Secondly, apperception inserts the *a priori* concept of the ether into the distributive unity of objective perceptions, resulting in a systematic unity of experience (absolute unity of consciousness) that corresponds to the systematic unity of the moving forces of matter (or the ether). Consequently, the transition establishes the correspondence between mind and world that Kant's post-Critical conception of physics requires.

Far from being separate issues, Kant's redescribed transition project and the gap problem are intimately connected and it is only by recognizing both the transcendental material and formal functions of the ether that the transition can be effected and the gap in Kant's Critical philosophy filled. The post-Critical theory that emerges in *Convoluts* 10–11 should not be seen as an abandonment of the Critical project. On the contrary, only Kant's post-Critical system is capable of bridging the gap within the Critical philosophy. In addition to filling the gap in the Critical philosophy, *Convoluts* 10–11 also serve to solve two other closely related problems within the framework of *CPR*. With regard to the appendix to the Transcendental Dialectic, Kant is able to establish the systematic unity of nature as well as the systematic unity of the subject's cognitions (absolute unity of consciousness) *a priori*. The harmonization [*Zusammenstimmung*] of the subjective and the objective solves the top-down/bottom-up problems since there is no *in principle* discontinuity between nature and the subject's cognition of nature. This coming together of the subjective systematic unity of cognitions and the objective systematic unity of nature establishes the "sufficient mark of empirical truth" and consequently safeguards Kant's theory of cognition in the Transcendental Analytic by dispelling the worries of the appendix to the Transcendental Dialectic. Establishing the sufficient mark of empirical truth in the appendix to the Transcendental Dialectic is intimately connected to the way in which Kant overcomes the dilemma that faces his theory of substance in the Analogies. Both require showing there is no *in principle* discontinuity between the systematic unity of nature and the subject's cognition (experience) of this systematic unity. Establishing the correspondence between them is sufficient for this task. Likewise, Kant's Ether Deduction in the *Übergang* section provides a *proof* for the totality of empirical reality (ether or Substance) that is only *presupposed* in the Transcendental Ideal of *CPR* but is *required* for the thoroughgoing determination of empirical objects (substances). In *Convoluts* 10–11, Kant completes his account of thoroughgoing determination by highlighting not only the transcendental material role that the ether plays in the thoroughgoing determination of empirical objects, but also the role that the *a priori* concept of the ether plays as the "formal principle" of thoroughgoing determination.

Notwithstanding all of the problems with the Critical philosophy that Kant's post-Critical transition project solves, one more problem remains, a problem which has plagued the Critical philosophy since its inception and threatens to undermine its very foundations. In the final chapter, we will examine how the results of Kant's post-Critical transition project can be applied to the problem of affection. Although recent commentators on *OP* tend to focus on the transition project itself rather than on the problem of affection *per se*, early commentators tended to ignore the transition project *per se* in favor of how *OP* might offer a solution to the problem of affection. I will argue that if one accepts the results of Kant's transition project, as I have articulated them, one will be in a position to solve the problem of affection. If I am right, then Kant's transition project should be considered the post-Critical culmination of his Critical philosophy.

NOTES

1. See *CPR* B160–161.
2. For Kant's comments on time, see *CPR* A35–36/B52.
3. See *CPR* A30–31/B46 and A32–33/B49–50.
4. See also *CPR* B233 and B257.
5. I am here following Caygill's interpretation of this passage. See Caygill, "The Force of Kant's Opus Postumum," 40.
6. Translation is my own. See also *OP* 22:367.
7. See *CPR* A374 where Kant draws this distinction.
8. For cognition requiring the unification of intuition and concept, see *CPR* A51/B75.
9. Adickes claims that the ether is imperceptible, though he argues that it is imperceptible since it lacks secondary qualities. For creatures like us, secondary qualities are required for the perception of any object. Secondary qualities are associated with complexes of sensation [*Empfindungskomplexe*]. Although the ether is the *source* of sensation, it is not itself a sensation and so secondary qualities cannot be attributed to it. See Adickes, *Kants Lehre von der doppelten Affection unseres Ich: als Schlüssel zu seiner Erkenntnistheorie* (Tübingen: Mohr, 1929), 70–74.
10. I will return to this idea in §4.3.
11. As Beiser puts it, 'Perceptions are the result of the forces acting upon it [the perceiver] and its reactions to them.' See Beiser, *German Idealism,* 187.
12. See *OP* 22:356–357, where Kant talks about the moving forces of matter as being necessary though not sufficient for objects of experience.
13. Translation is mine.
14. See *MFNS* 4:481.
15. Here I am following Félix Duque Pajuelo, "El problema del éter en la física del siglo XVIII y en el 'Opus postumum' de Kant," *Revista de Filosofía* 2, no.1 (1975): 42. For a discussion of Duque Pajuelo's work on *OP* (in German), see Basile, *Kants Opus postumum,* 350–352.
16. Edwards goes the other way, holding that *OP* turns space into something transcendentally real which conflicts with Kant's claims in the Transcendental Aesthetic. Edwards is led to this conclusion, however, since he considers the ether the 'subject-independent causal basis of all outer perceptions.' See Edwards, "Spinozism, Freedom, and Transcendental Dynamics," 65–67. As I argued in §3.4, however, Kant is not claiming that the ether is transcendentally real. In the next chapter, I will argue that the subject and the ether are not *independent* but are rather mutually *interdependent* for their determination.
17. Translation is mine. Kant makes similar claims in the *Übergang* section. For example, see *OP* 21:219, 224, and 229.
18. Translation is mine. Edwards suggests that the ether can be connected to time through the role it plays in the conservation of motion. See Edwards, "Spinozism, Freedom, and Transcendental Dynamics," 68.
19. Karl Ameriks argues that systematicity served two important purposes in the Modern period. It was acknowledged as the only honest way of meeting

skepticism and also allowed one to incorporate the advancements of modern science. See Karl Ameriks, "Kant on Science and Common Knowledge," in *Kant and the Sciences,* 41.

20. Translation is mine.
21. Translation is mine. Adickes also discusses the relationship between the unity of matter and the unity of experience in *Kants Opus postumum,* 244n1.
22. I will return to this issue later in this section when discussing the top-down/bottom-up problems.
23. Beiser and Caygill make a similar point. See Beiser, *German Idealism,* 203 and Caygill, "The Force of Kant's Opus Postumum," 35–36. Caygill ties this redefinition of "physics" to Kant's polemic against Newton's *Principia* in *Convolut* 11.
24. In particular, one might note the important role that the Analogies play in the Mechanics chapter of *MFNS.* For example, Kant's argument for the conservation of the quantity of matter in Proposition Two of the Mechanics assumes the soundness of Kant's argument in the First Analogy. For further discussion of this issue, see Friedman, *Kant's Construction of Nature,* 318. I called the First Analogy's argument into question, however, in Chapter One.
25. Translation is mine.
26. Translation is mine.
27. See *CPR* Bxvi–xviii, B128, and B137.
28. Translation is mine.
29. Bayerer, "Bemerkungen zu einem neurdings näher bekannt gewordenen Losen Blatt aus Kants Opus Postumum," 130 (lines 1–6). Translation is mine.
30. Translation is mine.
31. See *CPR* A161/B200.
32. Although I have argued that the function of the *a priori* concept of Substance is different from the function that the categories are supposed to serve within Kant's theoretical philosophy, one might still worry that the *a priori* concept of Substance calls into question the completeness of Kant's table of categories. In the B-edition Metaphysical Deduction, Kant claims not only that the table of categories is complete but that *MFNS* provides proof of this within 'a planned speculative science' (*CPR* B109–110). Does the *a priori* concept of Substance cast doubt on Kant's proof in *MFNS* and by extension on the completeness of the table of categories? As mentioned above (§1.1), the table of categories is derived from the logical table of judgments. If the table of categories is incomplete, presumably the corresponding logical table of judgments is incomplete as well. Kant himself, however, says things that suggest that the logical table of judgments might be incomplete. In *CPR,* when discussing disjunctive judgments within the logical table of judgments, Kant describes them simply as *exclusive* (*CPR* A74/B99). In *MFNS,* however, Kant recognizes both *alternative* and *distributive* disjunctive judgments different from the exclusive disjunction (*MFNS* 4:559–560). This suggests that incompleteness (either at the level of logical judgments or categories) need not be anathema to Kant. For defenses of the *completeness* of the tables, however, see Allison, *Kant's Transcendental Idealism,* 136–146 and Klaus Reich, *The Completeness of Kant's Table of Judgments,* trans. Jane Kneller and Michael Losansky (Stanford: Stanford University Press, 1992).
33. See *CPR* A224/B271. Here I am following Friedman, *Kant's Construction of Nature,* 29.
34. Kant discusses the Axioms, Anticipations, and Analogies at several other points of *Convoluts* 10–11 including 22:280, 22:292, and 22:345.
35. Adickes also compares the ether to this magnetic material. See *Kants Opus postumum,* 65.

36. Translation is mine. Kant says the same thing in the *Elementarsystem* which underscores the continuity in Kant's thinking concerning the modal status of the ether throughout the various stages of the transition project. See *OP* 22:188.
37. See *CPR* A226/B279.
38. See *OP* 22:518.
39. See *OP* 22:502.
40. See *OP* 22:437.
41. See *CPR* A160/B199.
42. Translation is mine. Throughout much of *CPR,* the most that can be expected of the understanding is a distributive unity of perceptions, but at one tantalizing point, Kant *does* claim the understanding is an 'absolute unity' (A67/B92). Kant's philosophical support for this claim, however, comes only in *OP.*
43. Adickes also recognizes the problem Kant has harmonizing consciousness and the moving forces of matter. See Adickes, *Kants Opus postumum,* 351.
44. Bold emphasis added.
45. Bold emphasis and translation of "*hineingelegt*" added. Although Kant thinks that the understanding creates the unity of nature through what it inserts *a priori,* it is not a creation out of whole cloth. In both the Critical and post-Critical periods, what is unified is ultimately given *a posteriori.*
46. See *CPR* B128.
47. See *CPR* A114. Although I am emphasizing the role transcendental apperception plays in producing the unity of nature, it is important not to overlook the role that imagination plays as the engine of synthesis. See *CPR* A78/B103. Of the different forms of imagination Kant discusses in *CPR,* transcendental imagination is most relevant to the present discussion. Although it belongs to sensibility, transcendental imagination synthesizes sensible manifolds (either pure or empirical) in accordance with the understanding's pure concepts. See *CPR* B151–152.
48. See *CPR* A647/B675.
49. Agreement of cognition with its object is Kant's nominal definition of truth.
50. Philip Kitcher suggests an explanation along similar lines. See Philip Kitcher, "Projecting the Order of Nature," in *Kant's Critique of Pure Reason: Critical Essays,* ed. Patricia Kitcher (Lanham: Rowman and Littlefield, 1998), 219–238. Friedman argues that the idea of the systematic unity of nature as a *whole* can only be regulative because we can never embed the totality of all empirical propositions into a completely interconnected system. He suggests that systematically unifying *parts* of this whole can be enough, however, to provide a sufficient mark of empirical truth. See Friedman, *Kant's Construction of Nature,* 549. If Kitcher's argument is sound, however, it would seem that any subset of empirical propositions will have truth-conditions that reach beyond that subset.
51. Since I am discussing this problem within the context of the Critical philosophy, my use of "appearance" should not be understood in terms of Kant's post-Critical distinction between direct and indirect appearance. One should rather think of "appearance" as referring to 'the undetermined object of an empirical intuition' (*CPR* A19/B33).
52. Robert Hanna raises these two worries. See Hanna, "Kant's Theory of Judgment," *The Stanford Encyclopedia of Philosophy,* http://plato.stanford.edu/archives/sum2011/entries/kant-judgment. These problems should not be confused with the problems that Friedman raises (see §0.4). Whereas Friedman's worries have to do with coordinating the different procedures of *MFNS* (constitutive through determining judgment) and *CJ* (regulative through reflective judgment), the top-down and bottom-up problems that I discuss are internal to Kant's theory of experience as presented in *CPR.*

53. See *CJ* 5:180. The transcendental principle required in *CJ* is that of purposiveness [*Zweckmäßigkeit*]. For more on this principle, see my discussion of Mathieu's position in the Introduction (§0.4). Mathieu believes that Kant's dissatisfaction with the principle precipitates the gap in his Critical philosophy. For reasons mentioned in the Introduction, however, I believe Mathieu fails to correctly identify the gap.
54. In the First Introduction to *CJ*, Kant holds that the purposiveness of nature (*CJ* 20:219) and its systematic unity (*CJ* 20:209) are only presupposed as regulative theses. Beiser claims that the gap in Kant's Critical philosophy stems from his failure to establish the systematic unity of nature as a constitutive thesis. The problem, according to Beiser, is that without establishing the systematic unity of nature there is no guarantee that empirical physics is a science since, in order to be a science, it must be systematically unified (*CPR* A832/B860). According to Beiser, Kant already recognizes this gap in *CPR* (citing *CPR* A651/B679) and tries to bridge it in *CJ* and later in *OP*. See Beiser, *German Idealism*, 182–185. Although I agree with Beiser that Kant needs to establish the systematic unity of nature as a constitutive thesis and that this issue is closely tied to the gap problem, I believe the problem Beiser identifies in the Transcendental Dialectic actually stems from a problem Kant has in the Transcendental Analytic (*viz.* in the Analogies). In addition, I do not think Kant recognized the problem Beiser identifies as constituting a gap in the Critical philosophy during the Critical period. Otherwise, why would he claim, in the Preface to *CJ*, that the Critical philosophy was complete? Unlike Beiser, Allison argues that Kant need only establish the systematic unity of nature as a regulative thesis in the sense that the dynamical principles of the Transcendental Analytic are "regulative" (i.e., concerning only the relations between objects of experience and not the construction of these objects). See Allison, "Is the *Critique of Judgment* Post-Critical?," in *The Reception of Kant's Critical Philosophy*, 82. Allison's interpretation depends on their being no difference between the way that Kant uses "regulative" in the Transcendental Analytic and the way he uses it in the Transcendental Dialectic. Kant makes clear, however, that there is a difference. Whereas the regulative principles from the Transcendental Analytic are 'still constitutive in regard to experience' once an object of experience is given, the regulative principles from the Transcendental Dialectic are only 'maxims of speculative reason, which rest solely on reason's speculative interest' (*CPR* A664–666/B692–694). Kant even gives reason's search for systematic unity in nature as an example of the latter, saying it 'is a legitimate and excellent regulative principle of reason, which, however, as such, goes much too far for experience or observation ever to catch up with it' (*CPR* A668/B696).
55. See *OP* 22:345.
56. Translation and bold emphasis is mine.
57. This is the way that some of the earliest commentators on *OP* interpreted Kant's position. For example, see my discussion of Kemp Smith in the Introduction (§0.5). One might view Choi's interpretation of *OP* as a more recent example insofar as Choi emphasizes the role of the subject's *activity* in producing existing things while minimizing the role of the subject's *passivity*. See Choi, *Selbstbewußtsein und Selbstanschauung*, 87. In contrast, Beiser argues that Kant continually resists 'transcendental solipsism' in *OP* and that the transition project can be viewed as an extension of his argument in the Refutation of Idealism. See Beiser, *German Idealism*, 190–194. As will become clear, I certainly agree with Beiser on this point.
58. Instead of claiming that apperception takes on too great of a role in *OP*, Edwards worries that the ether supplants apperception as the 'highest point' of transcendental philosophy (*CPR* B133). See Edwards, "Spinozism, Freedom,

and Transcendental Dynamics," 64. As I will argue in the next chapter, apperception (as a transcendental formal condition) and the ether (as a transcendental material condition) have coequal status within Kant's post-Critical conception of transcendental philosophy.

59. Translation is mine.
60. See *OP* 22:345.
61. The "form" mentioned in the above quote (*OP* 22:297) is equivalent, I believe, to the 'formal element of sensible representation' mentioned at the end of §4.2 (*OP* 22:300).
62. In *Convolut* 10, Kant claims that the moving forces of matter are not only required for the unity of spatiotemporal experience (*OP* 22:346–347) but also for the formation of bodies (*OP* 22:378).
63. For uses of "substances," see *OP* 22:327, 22:358, and 22:533. For uses of "Substance," see *OP* 22:457, 22:462, and 22:503.
64. Melnick also gives the microphysical example of how photons perish while electrons and positrons arise. This example might motivate the need for Substance more than Kant's example of the burning wood perishing while smoke and ash arise. Whereas in the former example it is difficult to conceive what substance might persist through the microphysical replacement change, in the latter example one could have recourse to some microphysical substance that persists through the burning of the wood. See Melnick, *Kant's Analogies of Experience,* 72.
65. Although there are many differences between Kant's transition project and his project in *MFNS,* there is a methodological similarity. One can fruitfully compare, I believe, the transformation of subjective perceptions into experience in *Convoluts* 10–11 with his transformation of appearances into experience in the Phenomenology chapter of *MFNS.* The starting point (subjective perception) is the same and various stages of synthesis are necessary to generate experience. Although Friedman does not discuss *OP* specifically in *Kant's Construction of Nature,* his reconstruction of the Phenomenology invites this comparison. See Friedman, *Kant's Construction of Nature,* 434–437.
66. See *OP* 22:384.
67. On this point, I am following Beiser, *German Idealism,* 189.

5 Kant and the Problem of Affection

Without the thing-in-itself one cannot enter the Kantian philosophy, but with the thing-in-itself one cannot remain.

—F.H. Jacobi [1]

5.0 INTRODUCTION

Building on Jacobi's oft-cited dictum, Hans Vaihinger, in the late 19th century, posed a now famous trilemma for Kant's theory of affection: (1) If things-in-themselves are the affecting objects, then one must apply the categories beyond the conditions of their application (space and time). (2) If one holds that appearances are the affecting objects, then one must hold that these appearances which are the effects of affection are themselves the causes of affection. (3) If one holds that things-in-themselves affect the noumenal subject in parallel with appearances affecting the phenomenal subject, then that which is a representation for the noumenal subject must serve as a causally efficacious thing-in-itself for the phenomenal subject's production of an empirical representation of the very same object (so-called "double affection").[2]

All three of the trilemma's horns have their own *prima facie* difficulties. The first seems to require that Kant violate his own constraints on the application of the categories in *CPR*. The categories (e.g., causation) cannot be applied beyond the sensible conditions necessary for their application, *viz.* appearances in *space and time.*[3] Since things-in-themselves, according to transcendental idealism, are not appearances in space and time, the category of causation cannot be applied to them. If any analysis of affection must deploy causal concepts, (1) does not appear to be a viable option.

The second horn seems to require that appearances be made both the cause and effect of affection. The first worry with this view is patently obvious: how can something be the *cause* of itself? If one views appearances as *particulars,* one is led quickly down the path toward empirical idealism (e.g., Berkeleyan phenomenalism). If these particulars are sensible ideas or even intentional objects, how can they be the cause of their own existence? It seems that they must be, as Berkeley would have it, causally inert and wholly dependent upon mental activity.

The third horn, called "double-affection," is perhaps the most infamous solution to the problem of affection. Double-affection is 'double trouble' according to Van Cleve, since it inherits problems from both the phenomenal affection as well as the noumenal affection views. With regard to the former, Van Cleve asks how the appearance could affect the subject since it is simply a creature of representation. With regard to the latter, how could the noumenal object affect the subject since the categories (particularly causation) only have application to phenomenal appearances?[4]

As mentioned in the Introduction (§0.5), as well as at the end of the last chapter, most early interest in *OP* was generated by those who thought that an argument for double affection could be located in *OP* and that this theory might hold the key to understanding Kant's theory of cognition. The first section of this chapter will examine the theory of double-affection as well as the textual support for such a view in *OP*. After discussing the problems, both textual and philosophical, that double-affection faces, I will attempt to develop a new solution to the problem of affection, one which has its roots in *CPR* but is only fully realized through Kant's post-Critical transition project.

In trying to overcome the above trilemma, most commentators have relied on the assumption that appearances are themselves *particulars*.[5] Although there is ample textual evidence for such an interpretation, Rae Langton breaks from these commentators by holding that appearances are *relations*. Notwithstanding our points of difference, both Langton and I agree that appearances are relations. In the second section of this chapter, I will discuss Langton's interpretation as well as some objections to her interpretation. The third section of this chapter will examine those passages in *CPR* that suggest that Kant considered appearances to be relations. Using these latter passages, I argue that appearances, within the context of Kant's Critical philosophy, are spatiotemporal causal relations. Affection is one relational element within this ordered set of spatiotemporal causal relations, whose specific function it is to reveal structural features of the affecting object to the receptive subject. All causal relations must be spatiotemporal, however, insofar as causal relations (whether affective or not) can only obtain within space and time.

Even admitting that appearances are relations, it is still right to wonder whether the affecting object has an existence independent of the relation it bears to the subject or not. If the former, the affecting object might still be a thing-in-itself which could subject Kant to the problems of the first horn. If the latter, the affecting object might simply be a creature of the mind which could subject Kant to the problems of the second horn. I will argue that Langton minimally adopts the former view and consequently inherits the problems facing the first horn of Vaihinger's trilemma. Even so, there is reason to think that Langton is actually committed to a version of the third horn, in which case her view would inherit more problems than just those facing the first horn.

The fourth section of this chapter examines Kant's Transcendental Deduction of the Categories from *CPR* where he argues that the object of representation and the subject of representation depend upon one another for their determination. By expanding on Kant's relational conception of appearances, the Transcendental Deduction shows how Kant might avoid Vaihinger's trilemma. Since neither phenomenal objects nor subjects are determined without one another and it is through the appearance relation that they are determined, these relations are *intrinsic* to the relata so related. As I will argue, this differs from Langton's interpretation which views appearances as *extrinsic* to the relata so related.[6] Whereas Langton views appearances as a species of phenomena understood as extrinsic relations between independently existing things-in-themselves, I view appearances as intrinsic relations between phenomenal objects and subjects. In other words, the phenomenal object is not itself a relation, but one of the relata of the intrinsic appearance relation.

Even if the Transcendental Deduction succeeds in showing that the subjects and objects of representation are undetermined outside of the relationship they bear to one another, it explains only the *organization* of these representations, by apperception, in accordance with the categories. It does not, however, explain the *source* of the representations organized into determined objects and subjects. Another way of putting this worry is to ask what the material analogue of the spontaneity of apperception is when it comes to the receptivity of subjects to be affected from without? One way of answering this question is in terms of double-affection, which denies the genuine receptivity of phenomenal subjects to be affected from without. The activity of the noumenal subject is the ultimate cause of both phenomenal objects and phenomenal subjects. Early proponents of double-affection turned to *OP* as providing support for this interpretation. I will argue, in the fifth section of this chapter, that Kant's post-Critical transition project in *OP* offers a solution to this problem that does not require appeal to noumenal causes while maintaining the receptivity and spontaneity of subjects in the phenomenal world.

In *Convolut* 7, which Kant began writing immediately after *Convoluts* 10–11 and finished in December of 1800, he takes certain ideas developed within the transition project and applies them to the problem of affection. As we saw in the last chapter, a distinction arises in *Convoluts* 10–11 between *direct* and *indirect* appearances. Although these two forms of appearance are *intrinsic* to their relata and involve causal as well as spatiotemporal aspects (*pace* the Critical picture outlined above), they operate at very different levels. Whereas direct appearances are intrinsic causal-dynamic relations between an undetermined phenomenal object (physical activity of the ether or Substance) and an undetermined phenomenal subject (mental activity of apperception), indirect appearances are intrinsic causal-mechanical relations between determined phenomenal objects (physical bodies or substances) and determined phenomenal subjects (embodied cognitive subjects).[7] At the level of indirect appearance, Kant can maintain the standard account of empirical affection familiar from *MFNS, viz.* that phenomenal objects (substances) affect sensibility through the locomotive relations they bear to embodied cognitive subjects in a three-dimensional spatial framework. Of course, as we know from the Transcendental Deduction in conjunction with *Convoluts* 10–11, indirect appearances are themselves the result of the subject organizing direct appearances into determined phenomenal objects. Even if Kant's Critical view of appearances remains unchanged at the level of *indirect* appearance, at the level of *direct* appearance Kant comes to recognize, in *Convolut* 7, that the unification of physical activity and mental activity is necessary to generate both determined phenomenal objects as well as determined phenomenal subjects. Key components of this picture are familiar from both the *Übergang* section and *Convoluts* 10–11. Consequently, Kant's view in *Convolut* 7 can easily be seen as a natural outgrowth of his post-Critical transition project.

The final section of this chapter will return to Vaihinger's trilemma. I believe that my view on appearances dispels Vaihinger's trilemma since it does not require affection by things-in-themselves nor does it require affection by appearances. Within the Critical philosophy, appearances are intrinsic relations between phenomenal particulars (physical bodies including embodied cognitive subjects). These particulars are neither things-in-themselves, since they are in space and time, nor are they appearances, since appearances are themselves spatiotemporal causal relations between these particulars. Put slightly differently, whereas the *appearings* are relations, what *appears* is particular. Even so, neither determined phenomenal objects nor subjects are possible without one another and it is through the appearance that each is possible. Kant's post-Critical view does nothing to change the intrinsic nature of the appearance relation or its causal and spatiotemporal aspects. At the same time, however, by distinguishing between *direct* appearances and *indirect* appearances, Kant emphasizes the important role that both the ether's physical activity and apperception's mental activity play on either side of this intrinsic relation, something which inextricably binds these two forms of activity to one another.

5.1 THE THEORY OF DOUBLE-AFFECTION

Although I will argue below that Kant's theory of affection requires only *one* phenomenal world of interdependent objects and subjects, there is at least one way in which Kant obviously does talk about *two* worlds of objects. Both God and the soul (or noumenal subjects) are, if they do exist, *positive* noumena, i.e., objects that would exist as *substantiae noumena* with full sets of intrinsic non-relational properties.[8] Although these objects are merely thinkable within the context of the theoretical philosophy (e.g., the agent-causal noumenal subject of the Third Antinomy), they are absolutely necessary for the practical philosophy.[9] Keeping this in mind, I would like to turn to the last horn of Vaihinger's trilemma, the so-called "double-affection" view.

Adickes is the most famous proponent of this position and holds that the thing-in-itself affects the I-in-itself [*Ich an sich*] through logical-teleological relations. Adickes states explicitly that by the "I-in-itself" he means something *transcendent* and not merely *transcendental.*[10] Consequently, he must mean something like the noumenal subject when he talks about the I-in-itself. In what follows, I will contrast this noumenal subject with the empirical or phenomenal subject. According to Adickes, the relations between the thing-in-itself and the noumenal subject are logical-teleological and not spatiotemporal since space and time are features of the phenomenal and not the noumenal world. Even so, according to Adickes, the noumenal subject *translates* [*übersetzt*] these logical-teleological relations into spatiotemporal relations between complexes of force and the empirical subject.[11] Adickes believes that the spatiotemporal relations that the noumenal subject generates correspond to the logical-teleological relations that obtain between the thing-in-itself (noumenal object) and the noumenal subject. He uses the analogy of a two-dimensional painting being translated into a three-dimensional representation.[12] For Adickes, there are really three levels of objects. There is the thing-in-itself (noumenal object) which affects the noumenal subject (noumenal affection). There is the complex of force (appearance in itself) which affects the empirical subject (phenomenal affection).[13] Finally, there is a determined object of experience (appearance of an appearance) which results from the empirical subject synthesizing the representations generated by this latter type of affection.[14]

One great advantage of double-affection is that it can make sense of those conflicting passages of *CPR*, some of which imply that the affecting object is phenomenal as well as those that imply the thing-in-itself is the affecting object.[15] This virtue is not lost on Kemp Smith, and following Adickes, he claims that double-affection helps to bring together the disparate views of two competing schools of Kant scholarship. One school (represented by Fischer and Schulze) postulates things-in-themselves as the affecting objects while denying the possibility of affection by phenomenal objects. The other school (represented by Maimon and Fichte) holds that the whole story of affection must stop with phenomenal objects.[16] Both schools seem to find strong textual support in *CPR*. Double-affection then appears a natural solution insofar as it can accommodate both views while showing how they supplement one another. Without things-in-themselves affecting noumenal subjects and the synthetic acts of those noumenal subjects, there would be no spatiotemporal causal-dynamic world through which phenomenal objects could affect phenomenal subjects. Without phenomenal objects affecting phenomenal subjects, however, there would be no *experience* of this spatiotemporal causal-dynamic world including both phenomenal objects and the embodied phenomenal subjects they affect.

Although double affection is a complex view, and one not made explicit in *CPR*, Adickes nevertheless thinks that it holds the key to understanding Kant's theory of cognition.[17] From the standpoint of the empirical subject, the world of appearances comes equipped with spatiotemporal relations between complexes of force and other empirical subjects. Although the individual subject can provide the form for such relations, the individual subject cannot produce the relations themselves (content). There must be objective grounds for these relations that explain how empirical subjects can share the same phenomenal world.[18] Adickes believes that these objective grounds can be found in the logical-teleological relations that obtain between things-in-themselves and noumenal subjects. Consequently, noumenal affection is required by Kant's theory of cognition.

Even though these logical-teleological relations serve as the objective grounds for spatiotemporal relations, since things-in-themselves are not spatiotemporal, they cannot affect empirical subjects and consequently cannot directly communicate these relations to empirical subjects.[19] Again, the noumenal subject and its synthetic functions are necessary to translate these logical-teleological relations into spatiotemporal relations between complexes of force. Since the subject cognizes the *locations* of affecting objects, Kant requires phenomenal affection by these complexes of force. Both noumenal affection between things-in-themselves and noumenal subjects as well as phenomenal affection between complexes of force and empirical subjects are necessary for Kant's theory of cognition.

There is ample textual evidence for the *phenomenal* side of double-affection in *OP*, i.e., the moving forces of the ether producing appearances in themselves (direct appearances) which the subject then synthesizes into the appearance of an appearance (indirect appearances). The omnipresent and sempiternal moving forces of the ether or Substance (direct appearances) as well as discrete spatiotemporal objects or substances (indirect appearances) are wholly phenomenal. I already offered some textual evidence for this view in the last chapter (§4.1), but there are several other locations where Kant makes similar claims.[20] There is a paucity of evidence, however, for the *noumenal* side of double-affection in *OP*. Adickes recognizes that, from the perspective of Kant's *theoretical* philosophy, the thing-in-itself is a mere thought object and can be nothing more than the objective correlate of the noumenal subject's activity. For example, Adickes quotes passages from *Convolut* 7 where Kant says that 'the object in itself (noumenon) is a mere thought object (*ens rationis*). . . . the material—the thing-in-itself—is = X, is the mere representation of one's own activity' (*OP* 22:36.3–4 and 22:37.10–12).[21] Notwithstanding the deflationary interpretation of the thing-in-itself that *Convolut* 7 seems to support, Adickes insists that Kant was *privately* convinced that things-in-themselves exist and stand in affective relations with noumenal subjects.[22] Even so, he admits that Kant's realism with regard to noumena is only *publicly* expressed in his practical philosophy.[23] When it comes to the role the noumenal subject has in producing the spatiotemporal relations between complexes of force and the phenomenal subject, Adickes again quotes a passage from *Convolut* 7:

> I am an object of myself and of my representations. That there is something else outside me is my own product. I make myself. Space cannot be perceived. (But neither can the moving force in space be perceived, insofar as it is represented as actual without a body which exercises it.) We make everything ourselves.
>
> (*OP* 22:82)[24]

Passages like this one lead Kemp Smith to claim that the noumenal subject is doing all of the work in Kant's post-Critical theory of affection. Given the textual evidence mentioned above, there is no reason to think that things-in-themselves (noumenal objects) have a meaningful role to play in determining how the noumenal subject generates the phenomenal world. Likewise, given the extensive role that the noumenal subject has to play in generating the phenomenal world, there is little room for the efficacy of phenomenal objects and subjects. For Kemp Smith, Kant's thoroughgoing subjectivism in *OP* leaves him with no way to carve out a realist position about the natural world.[25] Insofar as Kant's transition project is aimed at establishing a robust empirical realism about the natural world (as I argued in §4.1), his post-Critical theory of affection would seem to undermine that project. Kemp Smith's view is contrary to my second principle of interpretation (from §0.1), however, insofar as it makes Kant inconsistent with himself.

Perhaps more troubling, Kant's theory seems to hinge on the activity of something which is in principle unknowable from the perspective of his theoretical philosophy.[26]

This leads to a host of other problems. As mentioned above, Adickes believes that logical-teleological relations underpin spatiotemporal relations. It seems, however, that these two kinds of relations are incomparable, unlike a two-dimensional picture and its three-dimensional representation. With the pictorial analogy, at least both are *spatial*. There is no shared property, however, when it comes to logical-teleological and spatiotemporal relations. This is a problem that Adickes seems to recognize when he admits that the two kinds of relations are "incomparable" [*unvergleichbar*].[27] Adickes might try to overcome this criticism by holding that the noumenal subject is responsible for applying the same synthetic functions to these logical-teleological relations as the empirical subject applies to spatiotemporal relations. By this process, the noumenal subject *translates* these logical-teleological relations into spatiotemporal representations that the empirical subject can then synthesize. What synthetic functions could Adickes be referring to in the case of the noumenal subject's activity? It seems the only candidates would be the unschematized categories (i.e., the categories *prior* to their transcendental time-determination). Why should we think, however, that application of the *unschematized* categories could ever result in *spatiotemporal* relations?

Robert Paul Wolff has a similar worry. Double-affection seems to rely upon an *analogy* when discussing the transcendent activity of the noumenal subject. What guarantee do we have that the transcendent activity of the noumenal subject mirrors, in any way, the empirical activity of the phenomenal subject? Double-affection seems to commit Kant to a 'pre-established harmony of faculties.'[28] There is no way, even *in principle,* of knowing that the analogy holds since the noumenal process is in principle unknowable.[29] To claim knowledge of such a process would simply violate the constraints that Kant places on knowledge in *CPR*. Even if we could know this process, however, the only conceptual resources available to the noumenal subject would be the *unschematized* categories. Again, it is unclear how application of the unschematized categories through the synthetic activity of the noumenal subject could ever produce a *spatiotemporal* world. Adickes' view seems to violate my third principle of interpretation insofar it is philosophically implausible.

One response to the above worries is to claim they are predicated on a two-world interpretation of double-affection which entails, given our cognitive constitution, an unknowable *process* of noumenal affection. I believe there is a way in which Adickes can resist this interpretation. Adickes acknowledges that Vaihinger was the first to find evidence for double-affection in Kant's texts. Even so, Vaihinger does not think that the two forms of affection can be united with one another without contradiction since it would require the *same* object to be both a thing-in-itself (for the phenomenal subject) as well as a representation (for the noumenal subject).[30] How can the same item be both a representation as well as a thing-in-itself? Adickes responds by deploying a dual-aspect theory of objects seemingly anticipating Gerold Prauß' and later Allison's dual-aspect theories.[31] As he says:

> Thing-in-itself and appearance are not to be regarded as two different natures. They are equally real even though in another way they oppose one another as portrayal [*Abbild*] and that which is portrayed [*Urbild*].[32]

Although a portrait of someone is itself a different object than the person being portrayed in the portrait, Adickes' point seems to be that the thing-in-itself and the appearance are not *two* objects, but rather *one* object with two-aspects. Both aspects are equally real, though the appearance can be compared to a thing-in-itself in the way that a portrait of something can be compared to that which it portrays.[33] Adickes makes much the same point when it comes to the difference between the subject as it appears and the subject-in-itself: 'They are not two different natures, rather one and the same nature is, on the one hand, in itself and timeless appearing, on the other hand, in the form of time.'[34] Although one might think double-affection is committed to a two-world view, Adickes here seemingly defends a one-world view of double-affection. This one-world view could also hold the key to overcoming Vaihinger's criticism. Vaihinger seems to overlook the possibility that the *same* object could have two different aspects (thing-in-itself and thing-as-represented), and instead simply assumes that an object can either be only a thing-in-itself or only a representation (exclusive "or").

It is not entirely clear, however, whether Adickes is really defending the same view that Prauß and Allison defend more than half a century later. Whereas they take Kant's dual-aspect theory of objects to be an *epistemological* thesis, Adickes seems to be defending a *metaphysical* thesis since both aspects of the object are equally real.[35] It is not simply that the object (or subject) must be *considered* as having a non-spatiotemporal aspect, but rather that the object (or subject) really does *have* this non-spatiotemporal aspect. When it comes to the subject, as mentioned above, Adickes explicitly endorses a *transcendent* view of the I-in-itself. Likewise, Adickes believes Kant never doubted that the thing-in-itself had a *trans-subjective* existence.[36] Given that the categories can only be used for *thinking* but not *cognition* beyond the bounds of sense (a point I will return to below), however, it is unclear whether we could ever be in an epistemic position to claim that objects of subjects do *in fact* have a non-spatiotemporal aspect.[37]

Even if one assumes a more moderate version of Adickes' view in line with the epistemic interpretation, there are a number of objections that the view would still face. Paul Abela's problem with the dual-aspect theory is that it focuses on the formal features of intuition (space and time) to the exclusion of the empirical reference relation. Following Guyer, Abela holds that the forms of intuition necessarily represent the structure of the epistemic subject instead of the object of knowledge.[38] Van Cleve has two problems with the dual-aspect theory. The first is that the theory is implausible. Assume that I am typing with shoes on my feet. If you consider me apart from my shoes, am I barefoot? The answer seems to be "no." Just because you consider me apart from my shoes does not entail that I am not wearing them. Perhaps what is meant is that *if* I am wearing no shoes, *then* I would, under that assumption, be barefoot. This leads to the second problem. The conditional claim is a tautology. If things considered apart from our forms of sensibility means simply that they are non-spatial, this is the same as saying that *if* things had no spatial characteristics, *then* they would have no spatial characteristics.[39] Hanna has three criticisms of the view: (1) if we *know* that objects are phenomenal, why even insist on a noumenal aspect? (2) If objects are taken by us as both noumenal and phenomenal, then they must be properties of some *third* thing, neither phenomenal nor noumenal (on pain of attributing mutually exclusive properties to the same object). This third thing would unnecessarily bloat Kant's ontology. (3) There are many passages where Kant seems to talk explicitly about an ontologically independent noumenal object.[40] Hanna's second criticism echoes Vaihinger's original criticism and seems especially pressing if Adickes is taken as holding a metaphysical dual-aspect view. How could one and the same thing have, for example, *both* spatiotemporal and non-spatiotemporal properties? Consequently, under either dual-aspect interpretation (metaphysical or epistemological), Adickes' view seems philosophically implausible.

As might be inferred from Vaihinger's trilemma, the theory of double-affection is itself meant to avoid the problems associated with the first two horns, i.e., the *dilemma* that F.H. Jacobi so famously formulated and which opened this chapter: without the thing-in-itself one cannot enter the Kantian philosophy, but with the thing-in-itself one cannot remain inside it either. Without the thing-in-itself causing appearances there would be appearances without anything that appears, a position Kant himself rejects.[41] Consequently, without the thing-in-itself there would seem no way to enter Kant's philosophy. Once within his philosophy, however, there seems to be no way of understanding how the thing-in-itself could cause appearances since the schematized categories (including causation) can have application only *within* the world of appearances. Despite its many virtues, it is difficult to make double-affection work under either a one-world (dual-aspect) view or under a two-world view.

If one rejects the dual-aspect theory for the reasons given above, returning to a two-world theory of double-affection seems to inherit the problems that plague both horns of Jacobi's dilemma. In addition to Van Cleve's objections mentioned at the outset of this chapter, there are other related worries. Whereas affection through the thing-in-itself is uncognizable in principle since the process takes place in the noumenal realm, affection through appearances remains mind-dependent since the spatiotemporal phenomenal world is merely a representation of the noumenal self. In the following sections, I will develop a theory of affection which is wholly phenomenal. Since this theory will make appeal only to affection within the phenomenal world, it will avoid the worries that plague the double-affection view which is committed to an unintelligible process of noumenal affection. At the same time, however, I will argue that this theory can avoid the charge of Berkeleyan phenomenalism.

5.2 A RELATIONAL VIEW OF APPEARANCES?

As mentioned in the Introduction to this chapter, although the theory of affection I develop is importantly different from Langton's theory, it is nonetheless greatly indebted to Langton's work. Our most significant point of agreement is the conviction that Kantian appearances should be understood as *relations* rather than as *particulars*. This idea opens up new possibilities for dealing with the problem of affection, unanticipated by the three horns of Vaihinger's trilemma (as well as most interpretations in the secondary literature), which all seem to assume that appearances are particulars. In this section, I will present the key components of Langton's theory before posing some objections to it.

Langton's view rests on a fundamental distinction between things-in-themselves and phenomena. Whereas things-in-themselves are substances that have intrinsic properties, phenomena are the relational properties of these substances.[42] Substances, according to Langton, must be capable of a "lonely" existence, i.e., existing independently of their relations to any other things. Since Kant rejects the idea of a bare substratum, substances must have some intrinsic nature that is non-relational.[43] This would suggest that substances must have intrinsic non-relational properties as a prerequisite for their lonely existence. Since phenomena are things-in-themselves insofar as the latter are related to other things, there is only one world of substances and that is a world of things-in-themselves. Even so, there are two non-overlapping sets of properties that things-in-themselves possess: (1) relational properties and (2) intrinsic non-relational properties.[44] According to Langton, phenomena as relational properties of things-in-themselves are not reducible to the intrinsic non-relational properties of things-in-themselves.[45] These phenomena are not substances, however, since they are not capable of a lonely existence.

Langton believes that appearances are a subset of phenomena. Whereas phenomena are the relational properties of things-in-themselves generally, appearances are the relational properties of things-in-themselves insofar as the latter are related to human minds.[46] Things-in-themselves are related to human minds through the causal powers of the former. Human cognition is receptive, which is to say that humans can cognize objects only insofar as they are affected by objects through causal powers.[47] Consequently, humans can cognize only the relational properties of things-in-themselves insofar as they are affected by things-in-themselves. This does not result in cognition of the intrinsic properties of things-in-themselves, however, given the fact that phenomena (the relational properties of things-in-themselves) are not reducible to the intrinsic properties of things-in-themselves. Rather, it results in a kind of epistemic humility. Although humans can have knowledge of the relational properties of things-in-themselves via appearance, and can know *that* things-in-themselves have intrinsic properties, humans cannot know *what* these intrinsic properties are.[48]

Knowing that things-in-themselves are substances that have intrinsic properties, and that these substances affect us through their relational properties, requires deploying the categories of substance and causation. This might seem to raise the problem associated with the first horn of Vaihinger's trilemma, *viz.* that the categories must be applied beyond the conditions necessary for their application (space and time). Langton hopes to obviate the problem of applying the categories beyond the bounds of sense by deploying the *unschematized* categories (i.e., the category prior to its transcendental time-determination) to the thing-in-itself.[49] Kant himself suggests such a use of the categories in the Schematism section of *CPR:*

> Now if we leave aside a restricting condition, it may seem as if we amplify the previously limited concept; thus the categories in their pure

> significance, without any conditions of sensibility, should hold for things in general, as they are, instead of their schemata merely representing them how they appear, and they would therefore have a significance independent of all schemata and extending far beyond them.
>
> (*CPR* A146-A147/B186)

Allison raises a problem for Langton's use of the unschematized category of substance. Her use of the category, though minimal, still violates the *discursive* nature of human cognition which requires the *union* of concepts and sensible intuition.[50] Kant says as much throughout *CPR,* but to focus on the most famous of passages:

> Without sensibility no object would be given to us, and without understanding none would be thought. Thoughts without content are empty, intuitions without concepts are blind. It is thus just as necessary to make the mind's concepts sensible (i.e., to add an object to them in intuition) as it is to make its intuitions understandable (i.e., to bring them under concepts). Further, these two faculties or capacities cannot exchange their functions. The understanding is not capable of intuiting anything, and the senses are not capable of thinking anything. Only from their unification can cognition arise.
>
> (*CPR* A51/B75)

Although one can *think* of things-in-themselves through mere concepts, this is far from *cognizing* these things-in-themselves. According to Langton, we can cognize *that* things-in-themselves have intrinsic properties, with the humility consisting only in our inability to know *what* these properties are. If Allison is correct, Langton's view is still not humble enough.

Even though Langton's discussion focuses on the unschematized category of substance, I believe that her account also requires deploying the unschematized category of causation. The latter, however, is a characterization that Langton would almost certainly resist. For Langton, affection is simply a causal relation. Subjects are affected by things-in-themselves through their causal powers (relational properties) where these powers are irreducible to the intrinsic properties of things-in-themselves.[51] Since appearances are constituted by these causal powers, which are a subset of phenomena, and phenomena are in space and time, Langton would hold that there is no problem applying the *schematized* category of causation to these phenomena. It is important to note, however, that receptive subjects are not affected by *properties* (relational or otherwise) on Langton's account, but rather by the *substances* that have these properties. The thing-in-itself affects the subject and so must serve as one of the relata in the affection relation. Assuming that the affection relation is a causal relation, this would require extending the category of causation to things-in-themselves. Kant claims in *MFNS,* however, that a relation (e.g., the affection relation) can only be experienced if both relata are themselves objects of experience.[52] The thing-in-itself (*qua* substance with intrinsic non-relational properties) is not an object of experience and so any relation that it participates in could not be experienced according to this criterion.

Even so, one could deploy the unschematized category of causation which affords the ability to *think* though not *cognize* that things-in-themselves affect subjects through their causal powers. Perhaps Langton could admit this while claiming that her account requires only the *thinkability* and not the *cognition* of things-in-themselves with causal powers. This could take the sting out of the first horn of Vaihinger's trilemma if it is viewed as precluding cognition of things-in-themselves while admitting their thinkability. At the same time, however, I believe this interpretation of Langton would remove much of her position's force. One of Langton's primary goals is to make consistent what seems to be an inconsistent triad of Kantian claims: (1) Things-in-themselves exist. (2) Things-in-themselves are the causes of phenomenal appearances. (3) We have no knowledge of things-in-themselves.[53] She attempts to resolve the apparent inconsistency by claiming: 'We can know *that* there are things that have intrinsic properties without knowing *what* those properties are,' and a bit later on, 'We can know that a thing has certain causal powers without knowing what its intrinsic properties are.'[54] The triad is consistent since we can *know* (1) that things-in-themselves exist with intrinsic properties and (2) that things-in-themselves affect us through their causal powers, even if we are (3) *ignorant* of what the intrinsic properties of things-in-themselves are. Under the thinkability view, however, all she is entitled to claim is we can consistently *think* that things-in-themselves exist with intrinsic properties and we can consistently *think* that things-in-themselves have causal powers. Although the triad might still be consistent, the first two claims of the triad are significantly weaker than the ones mentioned above. Under this view, Kant could no longer claim 'that things in themselves exist' and 'that things in themselves are the causes of phenomenal appearances.' But these are just the metaphysical claims that Langton attributes to Kant and attempts to defend in her book! To give an example unrelated to Kant, there is a significant difference between one's book *being* published and consistently *thinking* of one's book being published. In my own case, I did the latter for far too long before getting around to the former.

Do Langton's claims really require *cognition* of things-in-themselves? Are the unschematized categories simply limited to their use in *thinking*? At points, Kant suggests that the unschematized categories can be used to generate knowledge claims, so perhaps Langton can make the claims that she wants to make while avoiding the constraints that the discursivity thesis places on cognition. Although Kant does allow the unschematized categories to be used in generating minimal knowledge claims beyond the bounds of sense, these claims are far weaker than the ones that Langton requires. When trying to understand what kind of claims the unschematized category of causation might be used to produce beyond the bounds of sense, one should note that the unschematized category of causation reflects the hypothetical form of judgment from the logical table of judgments, *viz.* that of ground to consequent.[55] When discursive creatures like us use these unschematized categories to make theoretical claims beyond the bounds of our sensible intuition, they have 'only a logical significance' and are merely 'functions of the understanding for concepts, but do not represent any object' (*CPR* A147/B186–187). In other words, when one deploys an unschematized category beyond the bounds of sense in a theoretical context, one does nothing more than signify the corresponding function of thought represented in the logical table of judgments. Knowledge of things-in-themselves as causes, however, would seem to require the representation of *objects* (*viz.* things-in-themselves) as causes, which is not simply the representation of the antecedent of a conditional judgment. It seems that Langton requires not merely the signification of the ground-consequent relation for which the unschematized category of causation would be sufficient, but rather cognition of the cause-effect relation for which the schematized category of causation is required.[56] Of course, since the schematized category cannot be applied to things-in-themselves, Langton cannot make use of it in this context. Even if the unschematized category of causation cannot provide cognition of any cause or *real* ground (at the level of objects), however, it can still signify a *logical* ground (at the level of concepts). The latter is a point I will return to again below (§5.4). The task now, however, is to develop a relational theory of appearances that can overcome the objections that Langton's view faces.

5.3 APPEARANCES AS INTRINSIC RELATIONS

Is there any reason to think that Kant viewed appearances as relations? In the Transcendental Aesthetic, he defines "appearance" as the 'undetermined object [*Gegenstand*] of an empirical intuition' (*CPR* A20/B34). Kant seems to make himself very clear here. In line with the received view, appearances *are* objects, i.e., particulars. This is not Kant's final word on the subject, however, and in a section of the Transcendental Aesthetic added in the B-edition, Kant says much more about appearances:

> For the confirmation of this theory of the ideality of outer as well as inner sense, thus of all objects of the senses, as mere appearances, this comment is especially useful: that everything in our cognition that belongs to intuition (with the exception, therefore, of the feeling of pleasure and displeasure and the will, which are not cognitions at all) contains nothing but mere relations, of places in one intuition (extension), alteration of places (motion), and laws in accordance with which this alteration is determined (moving forces). But what is present in the place, or what it produces in the things themselves besides the alteration of place, is not given through these relations. Now through mere relations no thing in

> itself is cognized; it is therefore right to judge that since nothing is given to us through outer sense except mere representations of relation, outer sense can also contain in its representation only the relation of an object to the subject, and not that which is internal to the object in itself.
> (*CPR* B66–67)[57]

From this passage, it seems that the appearances of outer sense are actually relations. Given the strength of Langton's own argument for this conclusion, I believe we should grant that this is so. But I primarily do so because, as it seems to me, the crucial question is rather this: what *kind* of relations are appearances? Langton seems emphatic, in particular, about certain ways in which she takes Kant's view to differ from Leibniz's. According to Langton, whereas Leibniz believes that appearances are reducible to the intrinsic non-relational properties of monads, Kant believes that appearances are irreducible to the intrinsic non-relational properties of things-in-themselves. Although Kant believes that things-in-themselves serve as the non-sensible *ground* of appearances, he disagrees with Leibniz, who thinks that things-in-themselves (in his case monads) are a *part* of appearances in the way that men are part of a crowd.[58] Kant believes the latter view rests upon a mistake. As he says in the Amphiboly:

> Leibniz took the appearances for things in themselves, thus for *intelligibilia,* i.e., objects of the pure understanding (although on account of the confusion of their representations he labeled them with the name of phenomena).
> (*CPR* A264/B320)[59]

Leibniz attempts to reduce appearances as relations to the intrinsic non-relational properties of monads as things-in-themselves.[60] So far I would simply note, then, as an obvious extension of my earlier criticism of Langton, following Allison, that despite these efforts to adhere to Kant's anti-Leibnizian stance, her position has both Kant and Leibniz embracing a positive approach to the notion of "noumena." That is to say, analogous to Leibniz's positing of monads, her position involves a view of Kantian things-in-themselves as entities to which one is entitled to make a positive ontological commitment, despite the fact that there is nothing about them, as they are "in themselves," of which we are able to form a positive conception. As Kant says:

> In the end, however, we have no insight into the possibility of such *noumena,* and the domain outside of the sphere of appearances is empty (for us), i.e., we have an understanding that extends farther than sensibility problematically, but no intuition, indeed not even the concept of a possible intuition, through which objects outside the field of sensibility could be given, and about which the understanding could be employed assertorically.
> (*CPR* A255/B310)

There must be such an assertoric commitment, for Langton as for Leibniz, precisely to the extent that one is entitled to assert the existence of appearances in the first place. The problem runs even deeper for Langton, however, insofar as appearances are a special subclass of phenomena. Whereas the latter are relations between things-in-themselves, the former are relations between things-in-themselves and human minds. Since the latter relatum of an appearance relation (as a subclass of phenomena) must be a substance in its own right according to Langton, she must mean "noumenal subject" when she talks of a "human mind" as being one of the relata in the appearance relation. Thus, while I have argued that Langton is committed to the first horn of Vaihinger's trilemma, this would seem to commit her to something like the third horn of the trilemma insofar as an appearance is a relation between a thing-in-itself (noumenal object) and a noumenal subject productive of phenomenal experience. If this is right, Langton might well inherit more problems than originally thought. Instead of positing only an uncognizable *object* of affection, as the first horn seems to do, double-affection must posit a whole uncognizable *process* of affection.

How then shall we think about Kantian phenomenal objects, consistently with a relational approach to the notion of appearances? The crux of my proposal turns on Kant's claim that the inner determinations of phenomenal objects are nothing but outer relations and that a phenomenal object is the sum total of relations.[61] Kant contrasts what would be a *substantia noumenon,* or object of the pure understanding, with a *substantia phenomenon,* or that phenomenal object which we cognize through sensible intuition. As he says in the Amphiboly:

> In an object of the pure understanding only that is internal that has no relation (as far as the existence is concerned) to anything that is different from it. The inner determinations of a *substantia phaenomenon* in space, on the contrary, are nothing but relations, and it is itself entirely a sum total of mere relations.
>
> (*CPR* A265/B321)[62]

In contrast to a *substantia noumenon,* there is nothing 'absolutely internal' in these relations. There are no intrinsic non-relational properties to be found, but rather only 'comparatively internal' (*CPR* A277/B333) properties where the latter turn out to consist of outer relations. For Kant, we have no other conception of absolutely internal determinations except that of '*thinking,* together with that which depends on it, the *feeling* of pleasure or displeasure, and *desire* or willing' (*MFNS* 4:544). Attributing these properties to phenomenal substance, however, would undermine Kant's claim, in *MFNS,* that matter is inert.[63] Consequently, we can only conceive of a phenomenal substance in terms of comparatively internal relations where these always consist of parts of matter external to one another. When distinguishing between phenomena and noumena, Kant even says a phenomenon is simply an 'object in a relation' (*CPR* B306).

Immediately after the above quoted passage from the Amphiboly, Kant says that 'we know substance in space only through forces that are efficacious in it' (*CPR* A265/B321). He describes these forces in dynamical terms (attraction and repulsion). This suggests that appearances are, at least in part, causal-dynamic relations. Since these forces are the only way in which the subject is acquainted with phenomenal substance, the passage also suggests that the subject is affected through these moving forces. Although this might describe the matter of appearance, Kant holds, much as he did in the Transcendental Aesthetic, that these appearances have an *a priori* form.[64] Space and time are the 'form of dynamical relations' (*CPR* A276/B332).[65]

As I will argue below (§5.5), there are ultimately *two* kinds of phenomenal objects, both of which stand in different relations with the subject. There is the ether which stands in causal-dynamic relations with the subject (direct appearance) as well as physical bodies or substances which stand in causal-mechanical relations with the subject (indirect appearance). Whereas Kant obliquely characterizes affection in terms of a causal-dynamic relation in the Amphiboly, he explicitly characterizes affection in terms of a causal-mechanical relation in *MFNS* (locomotion). In *OP*, the subject is affected by internally moving forces (ether) as well as locomotion (substances), where the latter is ultimately dependent upon the former. This theme is familiar from Kant's transition project and will play an important role in his post-Critical solution to the problem of affection.

Putting the Amphiboly section together with the other passages from the Critical period, a very different conception of appearances thus arises, one that stands in stark contrast both to the received view and to Langton's relational view of appearances. Namely, appearances are best understood on this view not as particulars, but rather as relations—but precisely as relations whereby the phenomenal *object* so related to the phenomenal subject possesses no intrinsic non-relational properties in addition to the relational properties in question. As Kant says, phenomenal objects contain nothing *absolutely internal* or non-relational, but rather consist solely of that which is only *comparatively internal* or relational. These relations possess both causal and spatiotemporal aspects.

These relations are *intrinsic* insofar as they are *essential* to the relata so related. Without the necessary possibility of cognitive subjects like us, objects could be neither spatial nor temporal since space and time are *a priori* forms of the subject's intuition.[66] Without causal relations, these subjects would have no *a posteriori* cognitive content since all that is present *a posteriori* to cognition through sensible intuition are the causal relations that obtain between affecting phenomenal objects and receptive cognitive subjects. Without such representational content, however, there would be no way for the subject to become conscious of itself through the unity of apperception that Kant introduces in the Transcendental Deduction of the Categories and which I will discuss in the next section.

Before moving on to the Transcendental Deduction, however, and the completion of my case for this alternative version of the relational approach to appearances, it is important to deal with a few objections that one might raise at this point: (1) A phenomenal substance is not a genuine substance. (2) The causal relation, as described in the Second Analogy, is insufficient for affection as I understand it. (3) There can be no necessary connection between distinct existences. (4) The view I am articulating commits Kant to Berkeleyan phenomenalism.

First Objection: A Phenomenal Substance Is not a Genuine Substance

Langton rejects the idea that phenomenal objects are genuine substances since they lack intrinsic non-relational properties and so are not capable of a lonely existence. She goes on to argue that relations (including causal relations) imply the existence of bearers that are themselves capable of a lonely existence. Assuming that affection is a causal relation, phenomenal objects can then not be the affecting objects, since they are not genuine substances and so are unable to bear causal relations.[67] Since Kant does talk about phenomenal substances and seems committed to empirical affection by these substances, it is important to respond to Langton's concerns.[68] As I have argued, the unschematized categories cannot offer cognition of either substances or their causal relations since this is the exclusive function of the schematized categories. To what, however, do the schematized categories apply? As mentioned in Chapter One (§1.3), Kant defines the schematized category of substance as 'the persistence of the real in time' (*CPR* A144/B183). Consequently, the schematized category of substance could have application only to phenomena since only phenomena possess the appropriate *temporal* characteristics for application of the schematized category. Kant's statement of the principle of the First Analogy in the A-edition, quoted at the beginning of Chapter One, supports this interpretation. To repeat, there he says that 'all appearances contain that which persists (substance) as the object itself, and that which can change as its mere determination, i.e., a way in which the object exists' (*CPR* A182). The object itself in this case is a *substance* which is characterized in terms of its *persistence* (temporal determination) and finally this substance is *contained* within the appearance. Nothing other than a phenomenal substance could serve all of these functions.

Second Objection: The Causal Relation Is Insufficient for the Affection Relation

Notwithstanding what Kant says in the First Analogy, one might be concerned that the doctrine of causation on offer in the Second Analogy seems ill equipped to account for affection as I understand it, since the Second Analogy has to do with a rule-governed temporal relation between events and not with any relation that might obtain between objects and subjects.[69] This is an important objection to any view that takes affection as involving a causal relation between objects and subjects including Langton's own view.

The event based interpretation of the Second Analogy, however, is not the only interpretation on offer. Taking into account the other two Analogies which have to do with substances and not events, Eric Watkins has argued that the Second Analogy should not be seen as offering a simple *event* based model of causation, but rather as ultimately offering a *substance* based model of causation.[70] Whereas the Third Analogy describes a two-way causal relation between substances, the Second Analogy describes just one direction of this two-way relation.[71] In the Second Analogy, the temporally determinate change of a substance's states require a real ground for this change lying outside of the substance whose states are changing. This real ground is found in another substance which changes the states of the first substance through the exercise of its intrinsic causal powers.[72] Even if one does not accept Watkins' interpretation of the Second Analogy, however, the Third Analogy, which deals with causal community, articulates the general kind of relationship I have argued obtains between phenomenal subjects and objects:

> The relation of substances in which the one [e.g., subject] contains determinations the ground of which is contained in the other [e.g., object] is the relation of influence, and, if the latter reciprocally contains the ground of the determination of the former, it is the relation of community or interaction.
>
> (*CPR* B257–258)[73]

Although this relationship (substances in space mutually determining one another's modes) is not strictly identical to the relationship that I have in mind, it helps to illustrate that the notion of mutual determination or reciprocal real grounds is not itself incoherent by Kant's lights. Just as the determination of a phenomenal object in space requires the synthetic activity of the subject, so too does the determination of a phenomenal subject in time (unity of consciousness over time) require representing these phenomenal objects in space. By "mutual determination" I do not mean mutual production (in the sense that the father begat the son who begat the father). This would be viciously circular.[74] Rather, what I mean is that neither phenomenal objects nor phenomenal subjects would possess certain essential properties absent the relationship they bear to one another. The object possesses its spatiotemporal and categorical properties by virtue of its relationship to the subject. The subject is the real ground for these determinations of objects. The subject is conscious of its identity over time by virtue of its relationship to objects. The objects are the real ground for this determination of the subject.

A good example of the relationship I have in mind can be found in Robert Boyle's discussion of the first lock and key. Put simply, he claims that without the existence of the key, the lock is not really a lock, but rather just a hunk of metal organized in a certain way. The same thing goes for the key.[75] Both the lock and the key are determined to their essential function by virtue of the relationship that they bear to one another.

Another example, one that does a good job of illustrating the relationship between the ether (as undetermined phenomenal object) and apperception (as undetermined phenomenal subject), is the relationship between points and line segments in the first postulate of Euclid's *Elements*. Whereas a straight line segment is uniquely determined by its two endpoints, these points are uniquely determined by the straight line that is drawn between them. Even so, a point is not reducible to a line or vice-versa. Although Euclid has a primitive definition of "point" as that which has no part and "line" as breadthless length, the geometrical objects that these terms denote are completely indeterminate without reference to one another. This is analogous to the intrinsic relationship between the ether and apperception where both are undetermined in themselves and mutually determine one another through their effects in the phenomenal world. I will return to this relationship in §5.5.

Causal community, as interpreted within the context of my theory of affection, is not "double affection" since only *one* kind of affection is being described, reflecting the interdependence of subjects and objects in the phenomenal world as the real grounds for each other, and not *two* different kinds of affection reflecting relations between subjects and objects in both a noumenal and a phenomenal world. This makes for a stark and fundamental contrast with Langton's view. Although appearances are *relations,* this does not entail that they are *extrinsic* or nonessential to the relata so related. And although appearances are constituted by *intrinsic* properties, this does not entail that they are *non-relational.* Appearances are *nothing but* a lawfully ordered set of spatiotemporal causal *intrinsic* relations. Neither the phenomenal object nor the subject related through appearances is determined without the other. This is not the case for Langton, however, when one considers appearances as a subset of phenomena in her sense. Phenomena as a class and appearances as a subclass of phenomena are *extrinsic* relational properties between things-in-themselves. In the former case, these are relations between things-in-themselves *generally,* whereas in the latter case, these are relations between things-in-themselves and noumenal subjects. These relations (either phenomena or appearances) are not *essential* to the relata so related, since the latter are substances with intrinsic non-relational properties and would exist independently of any relation to other things. Put slightly differently, whereas Langton believes that both phenomenal objects and appearances are merely relations, I believe that phenomenal objects are particulars which bear intrinsic appearance relations to phenomenal subjects.

As I will argue below (§5.5), Kant's account in the Third Analogy prefigures his account of affection in *Convolut* 7 of *OP* much as it prefigures his

Ether Deduction in *Übergang* 11. In *Convolut* 7, the ether comes back to play an important role in Kant's theory of affection. Although Kant's post-Critical theory of affection departs from the view presented in this section in important ways, what does not change is Kant's commitment to affection as taking place only within the phenomenal world. Post-Critically, both the physical activity of the ether and the mental activity of apperception are undetermined outside of the relationship that they bear to one another. It is only through this relationship that determined phenomenal objects and determined phenomenal subjects emerge. As I have already suggested, but will argue for in the next section, determined phenomenal objects and determined phenomenal subjects can also be viewed as intrinsically related within the context of the Critical philosophy.

As I argued in §1.3 and §4.2, the schematized categories have proper application only to substances. This does not entail, however, that the ether is not phenomenal. As omnipresent and sempiternal, it has both spatial and temporal properties (§4.1). Kant even comes right out and says, 'The first principle of representation of the moving forces of matter [is] to regard them not as things-in-themselves, but rather as phenomena according to the relations which they have to the subject' (*OP* 22:300). These relations are affective. The ether is an undetermined phenomenal object that affects the subject in sensibility (direct appearances) through its internally moving forces. Even though the ether itself cannot be subsumed under schematized categories as an object of experience, its effects (sensations) can be subsumed under these categories (generating indirect appearances). Furthermore, the *a priori* concept of the ether contains causal efficacy as one of its marks (§2.0), a concept which is itself applied through the schematized categories in their role as Principles (§4.2).

Third Objection: No Necessary Connection Between Distinct Existences

One might object to my position by deploying Hume's dictum that there can be no necessary connection between distinct existences.[76] If there can be no necessary connection between distinct existences, then how can subjects and objects be distinct existences, given that they are necessarily connected? I think the best way for Kant to avoid this objection is to deploy his distinction between material and logical necessity from the Third Postulate of Empirical Thought. Whereas the former concerns 'only the relations of appearances in accordance with the dynamical law of causality,' the latter concerns 'the connection of concepts' (*CPR* A226–227/B279–280). Although one cannot derive the existence of an object simply by examining the concept of a subject, just as one cannot derive the existence of a subject simply by examining the concept of an object, subjects and objects are nonetheless necessarily connected according to the dynamical law of causality as it operates in the appearance relation. This conception of necessity also fits well with Kant's discussion of causation in the Third Analogy, which, as I suggested above, is the best way, within the Critical period, of understanding the reciprocal relation that obtains between subjects and objects via the intrinsic appearance relation. If one views the Humean objection as directed at a logically necessary connection, then Kant can avoid the objection while still maintaining a materially necessary connection between subjects and objects. To put things somewhat differently, the relation between subjects and objects is one of *synthetic* not *analytic* necessity.

Fourth Objection: The View Commits Kant to Berkeleyan Phenomenalism

Returning to Vaihinger's trilemma, he claims that if one holds that appearances are the affecting objects, then one must also hold that these appearances serve as their own causes. I believe that this characterization of appearances relies on the assumption that appearances are particulars, where it is natural to ask how an object can be the cause of itself. Most commentators assume (either explicitly or implicitly) that appearances are particulars. Vaihinger's worry can be avoided, however, once one understands that the appearance is not a particular, but rather a relation. The appearance does not cause or depend wholly on itself. As an intrinsic relation, the appearance depends both upon the phenomenal object and phenomenal subject.

Although the distinction between appearance (relation) and phenomenal object (particular) plays an important role in avoiding the second horn, the intrinsic nature of the appearance relation also plays a role in avoiding this horn, insofar as the intrinsic nature of the appearance relation helps to clarify how Kant can avoid the Berkeleyan style idealism that the second horn seems to entail. For example, one might accept that the appearance is a relation while holding that the phenomenal object that enters into this relation is merely a virtual or intentional object. I believe that this simply returns us to Berkeleyan phenomenalism in a different guise, insofar as the phenomenal object would still depend entirely upon the mental activity of the subject to generate the object through acts of construction (virtual) or mental directedness (intentional). Under my view, however, the phenomenal object is not a mere intentional object or construction of sensible ideas, since the phenomenal subject is just as dependent upon the phenomenal object that affects it as the phenomenal object is dependent on the phenomenal subject that formally conditions it. For just this reason, the phenomenal object is not an independently existing thing-in-itself nor is it a mere intentional or virtual object. Instead of a one-way *dependency,* the phenomenal object and subject are *interdependent,* and this allows Kant to avoid *both* the first and second horns of Vaihinger's trilemma.

Perhaps there is a way, however, of understanding appearances as intrinsic relations within the context of Berkeley's system. Much as Kant claims that phenomenal subjects and objects are intrinsically related via the appearance relation, couldn't Berkeley claim that minds and sensible ideas are intrinsically related via the perception relation? If to be is to be perceived or to be a perceiver, wouldn't a perceiver (mind) need something that it perceives in order to exist much as something that is perceived (sensible idea) needs something that perceives it in order to exist?[77] I believe this surface similarity, however, betrays a more fundamental difference. For Berkeley, ideas are inert and cannot be the cause of anything at all.[78] For Kant, however, phenomenal objects are required for empirical affection. Whereas phenomenal objects and subjects are interdependent (two-way relation) under Kant's view, the mind is ontologically primary for Berkeley and produces or can produce ideas through its will (one-way relation).[79] For Berkeley, the mind does not exist simply through the understanding by which it perceives ideas, but also through the will by which it produces or can produce them. Thus while the Berkeleyan view under consideration interprets perception to be a relation analogous to the appearance relation, perception is perhaps best seen for Berkeley as rather an intentional relation. For Kant, however, appearances are spatiotemporal causal relations between phenomenal affecting objects and receptive cognitive subjects, where the former are therefore not merely intentional objects.

Although Kant's view of appearances as relations is incommensurate with Berkeley's view of the perception relation, if one chooses to interpret Kantian appearances as particulars, comparing them to Berkeleyan sensible ideas brings into stark relief the problem that the second horn of the trilemma faces. If appearances are mind-dependent particulars, then how are they different from Berkeleyan sensible ideas? If they are not different and sensible ideas are causally inert, then how could appearances be the causes of affection, much less the causes of themselves? As I have argued, viewing Kantian appearances as intrinsic relations allows one to overcome this problem and in a way that does not commit Kant to any form of Berkeleyan phenomenalism.

Van Cleve would still object to my interpretation of appearances as relations, by claiming they are all 'structure with no stuffing.'[80] This objection has some force insofar as the nature of phenomenal objects is exhausted by the relations through which they appear. Given the comparatively internal nature of phenomenal objects and the fact that these objects are nothing but relations, one might wonder how the phenomenal object could be different from the appearance relation. Isn't there a distinction to be drawn, however, between the *appearing* object and those *appearances* through which it *appears*?

I believe Van Cleve's mistake is to consider the relation "in itself" apart from its relata. Kant conceives of appearances, however, as *intrinsic* relations. Consequently, one cannot consider the relation (structure) apart from its relata (stuffing) since the relation is *essential* to both relata. In contrast, Van Cleve falsely assumes that all relations are extrinsic. Van Cleve's own "virtual object" view of appearances makes them empirically ideal. This is a category mistake since it confuses the appearance for a sensible idea (or at least a construction out of sense-data). This interpretation of the term "appearance" would be unwarranted within Kant's theory of cognition.

Even so, it seems almost paradoxical to say that the phenomenal object could serve as one of the relata in the appearance relation while at the same time being exhausted by these relations. To its credit, Langton's view does not face this paradox. Whereas I take appearances to be *relations* between *interdependent* phenomenal objects and subjects, she takes both appearances and phenomena to be *relations* between *independently* existing things-in-themselves. Once one recognizes the *intrinsic* nature of the appearance relation, however, this apparent paradox begins to resolve itself. To say that the phenomenal object is exhausted by relations is not to say that the phenomenal object is not an object (or relatum of the appearance relation), but rather that the phenomenal object cannot exist outside of these relations. Failure to recognize the intrinsic nature of the appearance relation is what leads philosophers to search for relata beyond the bounds of sense. Once one acknowledges, however, that the relata of the appearance relation cannot exist beyond the bounds of sense, one need not go beyond the phenomenal world in order to find the relata of the relation. Even though I think recognizing the intrinsic nature of appearances is important for overcoming the paradox, Kant is only fully able to dispel the air of paradox post-Critically when he recognizes that the ether (as undetermined phenomenal object) and apperception (as undetermined phenomenal subject) are the ultimate relata of intrinsic appearance relations. We will return to this idea below (§5.5).

So to summarize: on my interpretation, appearances are not to be taken in themselves as *particulars,* but are rather intrinsic *relations* where one must always give an account of the relata for which this relation is essential, *viz.* phenomenal objects and subjects. What remains to be shown, however, is that neither the affecting object of representation nor the cognitive subject of representation really is possible outside of the appearance relation, or that most fundamental way in which phenomenal objects and subjects are related. Although I have given some reason for thinking that Kant might subscribe to this view, the Transcendental Deduction of the Categories makes clear that he does subscribe to this view by giving an *argument* for it.

5.4 HOW OBJECTS AND SUBJECTS MAKE ONE ANOTHER POSSIBLE

Even admitting that appearances are relations, it is still right to wonder whether Kant has yet provided an argument for why these relations must be *intrinsic.* Can the affecting object have an existence independent of the relation it bears to the subject or not? If the former, the affecting object might still be a thing-in-itself which could subject Kant to the problems facing the first horn. If the latter, the affecting object might simply be a creature of the mind which could subject Kant to the problems facing the second horn. I have given some reasons above for how Kant can avoid these horns, but his most elegant argument against both positions comes in the Transcendental Deduction of the Categories.

In the Transcendental Deduction, Kant claims that appearances are the means by which subjects are immediately related to objects in empirical intuition, but such objects of representation are undetermined unless these representations are united in the concept of what it is to be an object (categories).[81] The unification of representations requires synthesis and consequently the unity of consciousness (apperception) in the synthesis of them.[82] Consequently, according to Kant, 'the unity which the object makes necessary can be nothing else than the formal unity of consciousness in the synthesis of the manifold of representations' (*CPR* A105).[83] For these unified representations, we think of some object of representation as corresponding to them. This is the "transcendental object = X" which is always one and the same thing: a transcendental placeholder for the object of representation. The concept of a transcendental object is necessary, however, so that there is something to which cognition is related. As Kant says:

> The pure concept of this transcendental object (which in all of our cognitions is really always one and the same = X) is that which in all of our empirical concepts in general can provide relation to an object, i.e., objective reality. Now this concept cannot contain any determinate intuition at all, and therefore concerns nothing but that unity which must be encountered in a manifold of cognition insofar as it stands in relation to an object. This relation, however, is nothing other than the necessary unity of consciousness.
>
> (*CPR* A109)

At this point, it might appear that Kant's view on the object of representation is much in line with something like Van Cleve's "virtual object" interpretation. It is important to point out, however, that although the determined object of representation is possible only through the subject's activity, the determined subject of representation is itself possible only through the objects of representation. As suggested above, subjects and objects of representation make one another possible by mutually determining one another. Although I have already discussed how the subject of representation determines the object of representation, in the Transcendental Deduction, Kant suggests that the reverse holds true as well. Without the synthetic unity of these representations (i.e., the determined object of representation), one would not be able to represent oneself as an identical subject enduring throughout these representations. As Kant says:

> Therefore it is only because I can combine a manifold of given representations *in one consciousness* that it is possible for me to represent the

> *identity of the consciousness in these representations* itself, i.e., the *analytical* unity of apperception is only possible under the presupposition of some *synthetic* one.
>
> (*CPR* B133)

Consequently, the subject of representation makes objects of representation possible by unifying these representations, but the synthetic unity of these representations make the subject of representation possible.[84] Since the subject of representation makes the object of representation possible, the object cannot be a thing-in-itself. Contrary to Langton, it is certainly not an independently existing positive noumenon equipped with intrinsic non-relational properties. Kant maintains his transcendental idealism with regard to all objects of representation given that they are nothing once one abandons the perspective of the subject of representation. Since the objects of representation make the subject of representation possible, the objects of representation cannot be mere creatures of the mind as the second horn of the trilemma would have it. Kant maintains his empirical realism with regard to all objects of representation. The object of representation is not a mere virtual object insofar as the subject of representation is possible only through the objects of representation.

All actual objects in the sense described above are determined phenomenal objects (*substantiae phenomena*), which is simply to say they are objects of experience or what I have called "substances." Given the nature of sensible intuition as well as the results of the Refutation of Idealism, the actuality of these objects is established non-inferentially. Kant holds that sensible intuition presents its object immediately.[85] Likewise, in the Refutation, Kant argues that we have immediate awareness of the existence of objects in space since this is necessary for the temporal-determination of consciousness.[86] In this way, the Refutation is also an argument for how determined phenomenal objects make determined phenomenal subjects possible.

One can connect Kant's argument in the Refutation to his causal-mechanical theory of affection in *MFNS* (and as developed in the other works discussed in §1.2) in a way that illustrates the intrinsic relation between determined phenomenal objects and subjects as mutually determining real grounds. The embodied cognitive subject occupies a relative space that serves as the center of a three-dimensional framework within which determined phenomenal objects are precisely located and affect the subject through their locomotion. According to the Refutation, without the embodied cognitive subject perceiving these objects, the former could not be determined as a *subject,* i.e., aware of its identity over time.[87] The embodied cognitive subject is likewise essential for the determination of these phenomenal objects. Without it being *embodied,* there would be no center for this three-dimensional framework relative to which phenomenal objects are precisely determined as occupying discrete regions in space. Without it being *cognitive,* these objects could not be categorically determined (e.g., with respect to their specific locomotion). Finally, without the necessary possibility of its *subjective* forms of intuition, there would be no three-dimensional framework at all!

According to my account, positive noumena (*substantiae noumena*) may or may not exist, but creatures like us will never know one way or the other, given the fact that we do not possess intellectual intuition. It must be conceded, however, that there are a number of places in *CPR* where Kant seems to refer to the thing-in-itself as the affecting object.[88] There are two ways in which interpreters approach the thing-in-itself in this context. As Graham Bird puts it, although "traditionalists" see Kant's commitment to the thing-in-itself as problematic, "revolutionaries" deny or minimize such commitments.[89] Whereas Langton's interpretation of the thing-in-itself falls into this traditionalist camp, Bird, for example, defends the revolutionary option. In any case, it is also important to remember that Langton's account requires *positively* noumenal things-in-themselves for affection. At one point, Kant seems plainly to reject such a view of the affecting object on pain of empirical idealism:

> For in fact if one regards outer appearances as representations that are effected in us by their objects, as things in themselves found outside us, then it is hard to see how their existence could be cognized in any way other than by an inference from effect to cause, in which case it must always remain doubtful whether the cause is in us or outside us.
>
> (*CPR* A372)

Undeniably, there are a number of passages where Kant seems to take the thing-in-itself as some sort of ground of appearance. He even goes so far as to characterize the relationship between thing-in-itself and appearance as embodying a conceptual or analytic truth. Although there are many examples, perhaps the best comes in the Preface to the B-edition of *CPR:*

> Yet the reservation must also be well noted, that even if we cannot *cognize* these same objects as things in themselves, we at least must be able to *think* them as things in themselves. For otherwise there would follow the absurd proposition that there is an appearance without anything that appears.
>
> (*CPR* Bxxvi-xxvii)[90]

But what should we infer from this conceptual truth? Must the object possess some positive nature beyond what appears to the subject? Although this passage is operating squarely within a one-world context (*viz.* 'same objects as things-in-themselves'), it might be argued, as Adickes seems to do (see §5.1), that Kant is committed to some transcendentally real aspect of the object in question, as relevant to its ability to appear, by way of affection, to a subject. It is important to note, however, the contrast between "thinking" and "cognition" in this passage. Kant's point, I believe, is that we must make use of the *concept* of something that logically grounds appearance, even if there is *nothing* to which this concept refers. As Kant says in the footnote to the above passage:

> But I can *think* whatever I like, as long as I do not contradict myself, i.e., as long as my concept is a possible thought, even if I cannot give any assurance whether or not there is a corresponding object somewhere within the sum total of all possibilities.
>
> (*CPR* Bxxvi)

In line with this, one might still wish to interpret this passage as only claiming that no distinct transcendentally real *object* need correspond to the concept in question, while holding that it is analytic that some transcendentally real *property* of an otherwise empirically real object corresponds to it.[91] One would reasonably suppose, however, that Kant's *skepticism* concerning any object corresponding to the concept should equally count against any *affirmation* of some transcendentally real property as corresponding to it. In any case, a more reasonable explanation is that passages of this sort suggest that, at least within the context of talk about affection, there need not be any transcendentally real object or aspect of an otherwise empirically real object that corresponds to our concept of a thing-in-itself. After all, in the Third Postulate, which deals with necessity, Kant claims that 'the necessity of existence can thus never be cognized from concepts . . . the mark of necessity in existence does not reach beyond the field of possible experience' (*CPR* A227/B280). According to Kant, mere conceptual truths (analytic) cannot have existential import.[92]

This is why I used the locution "logical ground" above (end of §5.2), instead of "real ground" (or cause), to describe the relationship between the thing-in-itself and appearance. Whereas a *real* ground would require *something* doing the causing, a *logical* ground can be merely *conceptual*. The logical ground of appearance is not a transcendentally real thing or aspect of a thing, I therefore suggest, but rather something merely conceptual. This concept, I will argue, must be thought as corresponding to the phenomenal object as a synthesized collection of appearances. How then should we understand this concept within the context of Kant's theory of affection? My suggestion is that we return to Kant's conception of the transcendental object = X, which, as noted, plays an important role in the Transcendental Deduction, insofar as it serves as a transcendental placeholder for the object of representation. In particular, even though the transcendental object is only 'the concept of something in general' (*CPR* A251) which denotes nothing in particular, when the synthesized manifold of appearances is related to this concept, the relation yields cognition of an empirical object.[93] Although Kant also characterizes the transcendental object as the cause of appearance in the Amphiboly, given its purely conceptual status, the term "cause" cannot signify anything more than the (logical) ground of a ground-consequent relation:

> The understanding . . . thinks of an object in itself, but only as a transcendental object, which is the cause of appearance (thus not itself appearance), and that cannot be thought of either as magnitude or as reality or as substance, etc. (since these concepts always require sensible forms in which they determine an object).
>
> (*CPR* A288/B344)[94]

Even though we must possess the concept of a transcendental object in order to logically ground appearance, this does not entail that there is *anything* corresponding to this concept. This interpretation is consistent with the above quote from the Amphiboly, where Kant claims that that the transcendental object cannot be thought of as possessing reality (e.g., as the transcendentally real aspect of an otherwise empirical real object) or being a substance (e.g., a transcendentally real object possessing intrinsic non-relational properties). Conjoining this view from the Amphiboly with Kant's discussion in the Transcendental Deduction, it then seems that the transcendental object serves two important functions: it is (1) the logical ground of appearance and (2) the transcendental placeholder for the object of representation or phenomenal object. As we will see in the next section, Kant continues, in his post-Critical philosophy, to identify the thing-in-itself with the transcendental object within the context of affection.

Earlier I discussed the differences between phenomenal objects (*substantiae phenomena*) and positive noumena (*substantiae noumena*). Although they are distinct kinds of objects, both phenomena and positive noumena are nevertheless viewed *substantively*. In contrast, the transcendental object is viewed as merely *conceptual*. Given this fact, is there any place for the transcendental object on either side of the phenomena/noumena divide? Immediately after the above passage from the Amphiboly, Kant goes on to explain how this transcendental object can be considered *noumenal:*

> If we want to call this object a noumenon because the representation of it is nothing sensible, we are free to do so. But since we cannot apply any of our concepts of the understanding to it, this representation still remains empty for us, and serves for nothing but to designate the boundaries of our sensible cognition.
>
> (*CPR* A288/B345)

Following Allison, one can understand the concept of a transcendental object = X from the A-edition Transcendental Deduction as the concept of a *negative* noumenon that Kant explicitly introduces in the B-edition of *CPR*.[95] Both should be distinguished, therefore, from positive noumena or *substantiae noumena,* which have full sets of intrinsic non-relational properties. Returning to the former, however, much as the transcendental object is something to which the understanding relates to appearances but which is not itself an appearance, so too is the negative noumenon something to which appearances are related in intuition but which is not itself an appearance. Although Kant does not mention "negative noumenon" explicitly in the A-edition of *CPR*, there is a passage that suggests identifying the transcendental object with the negative noumenon:

> All our representations are in fact related to some object through the understanding, and, since appearances are nothing but representations, the understanding thus relates them to a *something*, as the object of sensible intuition: but this something is to that extent only the transcendental object. . . . it also follows naturally from the concept of an appearance in general that something must correspond to it which is not in itself appearance. . . . Now from this arises the concept of a *noumenon*, which, however, is not at all positive and does not signify a determinate cognition of any sort of thing, but rather only the thinking of something general, in which I abstract from all form of sensible intuition.[96]

Here one must focus on Kant's claim that the concept of a noumenon with which he is concerned in this passage is 'not at all positive.' Thus he seems to be drawing a contrast, which will only become explicit in the B-edition, between negative and positive noumena.[97] Whereas the former is simply the concept of something in general and so equivalent to the concept of a transcendental object, the latter possesses its own positive nature that could only be cognized by a creature with intellectual intuition. Although Kant characterizes both the transcendental object and the negative noumenon as *causes* of appearance at different points, if what I have said above is correct, either the transcendental object or the negative noumenon should be understood as the logical ground of appearances rather than as bearing a causal relation to appearances.[98]

This is where my position differs markedly from Allison's position. Allison claims that the transcendental object (for him the empirical object taken under a non-empirical description) is what *causes* appearances and as such cannot be taken as spatiotemporal, on pain of assigning to the cause of appearance those features that only appearances themselves can possess.[99] In turn, Allison rejects the idea that the transcendental object is merely a *logical* ground (conceptual) of appearances on pain of Berkeleyan phenomenalism, since under this view there remains nothing beyond the appearances 'to be considered as it is in itself.'[100] By serving simply as a transcendental placeholder, however, I do not believe that the transcendental object should be taken as the non-spatiotemporal *cause* of appearances. Allison seems to conclude from the fact that, insofar as it affects the subject, the affecting object cannot be wholly *dependent* upon the subject and its epistemic conditions, that the affecting object must be in some way *independent* of the subject and its epistemic conditions. This overlooks the possibility, however, that the affecting object and the subject are *interdependent*. As I have argued, the latter is Kant's position and follows from his view that appearances are nothing but *intrinsic* relations between phenomenal objects and subjects. And as I have also argued, this does not result in any form of Berkeleyan phenomenalism precisely because the subject is itself determined through the relations it bears to phenomenal objects, i.e., through the way in which it is affected by these objects.

Although it is correct to say that the subject is affected by these objects, one must be careful in describing the nature of the object in question. In an important way, determined phenomenal objects (and so, consequently, determined phenomenal subjects) do not exist prior to the activity of apperception organizing representational content in accordance with the categories. What is the cause of the representations that apperception synthesizes into objects of representation and through which the subject of representation is possible? Put simply, what is the corollary of apperception's activity on the side of the phenomenal object? One cannot say that the *determined* phenomenal object is the cause of the representational content that apperception organizes since this representational content must itself be organized before there is a determined phenomenal object.

Here it is useful to recall Kant's object-oriented definition of "appearance" as the 'undetermined object of empirical intuition.' As I will argue in the next section, the ether is the *undetermined* object of empirical intuition that affects us in sensibility and so provides the representations that apperception synthesizes into the *determined* phenomenal objects necessary for determined phenomenal subjects. Although the ether might be the ultimate dynamical cause of our *representations*, we *experience* only determined phenomenal objects that stand in spatiotemporal causal-mechanical relations with us as determined phenomenal subjects. In other words, both the objects of experience (determined phenomenal objects) and subjects of experience (determined phenomenal subjects) *emerge* from the joint activity of the ether (undetermined phenomenal object) and apperception (undetermined phenomenal subject). Just as I have argued that neither determined phenomenal objects nor determined phenomenal subjects are possible outside of the relationship they bear to one another, I will argue that neither the undetermined phenomenal object nor the undetermined phenomenal subject is possible outside of the relationship they bear to one another. Consequently, the appearance relation is intrinsic *all the way down*.

5.5 KANT'S THEORY OF AFFECTION IN *CONVOLUT* 7

Although I have furnished a philosophical rational for double-affection, while also offering some criticisms of this view (see §5.1), I would like to take a closer look at the textual evidence to determine whether Kant subscribed to double-affection in *OP* as the proponents of double-affection maintain. At certain points in *Convolut* 7, Kant claims that apperception posits the manifold of intuition *a priori,* including the object that affects the subject as well as the subject that is affected.[101] As mentioned above, Kant even goes so far as to say that 'we make everything ourselves.' A bit later on he claims transcendental idealism is the doctrine that 'the world is only in me (transcendental *idealism*)' (*OP* 22:97). These passages certainly seem consistent with at least one aspect of the double-affection view, *viz.* that the spatiotemporal world is produced by the synthetic acts of the noumenal subject.

There is an overwhelming amount of textual evidence, however, that seems to count against the double-affection interpretation of *Convolut* 7. It is important to note that Kant never uses the phrase "I-in-itself" [*Ich an sich*], or "self-in-itself" [*selbst an sich*], and though he sometimes refers to the subject as a "thing-in-itself" [*Sache* or *Ding an sich*] this should not be understood as referring to a positively noumenal subject. When discussing the relationship between the subject and the thing-in-itself, Kant says:

> The subject is here the thing in itself because it contains spontaneity. Appearance is receptivity. The thing in itself is not another object, but another mode of making oneself into an object. The intelligible object is not an *objectum noumenon,* but the *act* of the understanding which *makes* the object of sensible intuition into a mere phenomenon.
>
> (*OP* 22:414–415)

This suggests that the subject, as thing-in-itself, should not be considered a *positive* noumenon, or an object with intrinsic non-relational properties. It is rather characterized simply in terms of its spontaneous activity, an activity operative only within the phenomenal world. As I will argue below, inferring from this activity to some sort of positively noumenal subject that underlies this activity would be to make the very mistake that Kant accuses Descartes of making in the Paralogisms chapter of *CPR*. In fact, Kant compares the subject's activity to the transcendental object = X, saying that this 'act of the understanding = X' (*OP* 22:94.10).[102] Kant says much the same thing about the object as thing-in-itself. Although there is one point at which Kant seems to mention the thing-in-itself (noumenon) as separate from the transcendental object (phenomenon), there are many other points where Kant seems to identify the thing-in-itself with the transcendental object = X.[103] At one point Kant says:

> The thing in itself is a thought-object (*ens rationis*) of the connection of this manifold whole into the unity to which the subject constitutes itself. The object in itself = X is the sense-object *in itself,* but as another mode of representation, not as another object.
>
> (*OP* 22:414)

This passage might seem to support Adickes' dual-aspect theory of objects insofar as the thing-in-itself is not a different object from the sensible object, but rather is simply another way of representing the same object. Notice the locution "object-in-itself = X." Kant uses this phrase often in *Convolut* 7, sometimes substituting it with "thing-in-itself = X."[104] To my mind, these formulations suggest Kant's conception of the transcendental object = X, which plays such an important role in the Transcendental Deduction of the Categories. This similarity is not lost on Adickes and at one point he even identifies the thing-in-itself with the transcendental object.[105] It must be remembered, however, that the transcendental object = X is neither an independently existing non-spatiotemporal *object* nor the transcendentally real *aspect* of an otherwise empirically real object, but is rather simply the transcendental *placeholder* for the object of representation. As I argued in the last section, the thing-in-itself, within the context of the Transcendental Deduction, functions as nothing more than a transcendental object = X or negative noumenon. In *Convolut* 7, Kant seems to compare the thing-in-itself = X to something like a negative noumenon: 'The thing-in-itself = X is not another object but only another [standpoint], namely the negative standpoint from which we consider one and the same object' (*OP* 22:42.5–7).[106] A negative noumenon is simply the concept of a thing after we abstract away from the way in which we intuit that thing. Much as with the transcendental object, without any content from empirical intuition, the negative noumenon must always be one and the same thing = X, *viz.* a transcendental placeholder for the object of representation. In *Convolut* 7, Kant reiterates the point, familiar from the Transcendental Deduction, that this placeholder is necessary so that there is something to which cognition is *related*.[107] In its role as placeholder, the transcendental object = X serves as the locus for the subject's combination of the manifold of representations into the cognition of a determined object of experience.[108]

In *Convolut* 7, Kant does not adopt a two-world theory of objects or subjects. Consequently, any theory of double-affection that requires two-worlds would seem to be precluded by Kant's discussion in *Convolut* 7. When talking about the "thing-in-itself," Kant makes clear that this term does not denote a *positive* noumenon, but at most a *negative* noumenon or transcendental object. Although Kant does talk about objects as having two aspects in *Convolut* 7, he does not talk about these objects as having two sets of properties. Under the most charitable interpretation, Adickes' theory seems to require that one and the same object have different and seemingly incompatible properties. In their noumenal aspect, objects must be non-spatiotemporal and must bear logical-teleological relations to the noumenal aspect of subjects. In their phenomenal aspect, objects must bear spatiotemporal causal-dynamic relations to the phenomenal aspect of subjects. According to Kant's account in *Convolut* 7, however, the object has only *phenomenal* properties and the transcendental object is simply the concept of this object stripped of all these properties. Again, the transcendental object is a mere thought object and neither a transcendentally real thing nor a transcendentally real aspect of an otherwise empirically real thing.

Assuming that the thing-in-itself functions as a mere thought object in *Convolut* 7, is there any way of avoiding the second horn of Vaihinger's trilemma? If the thing-in-itself is stripped of its transcendental reality, can phenomenal objects be anything more than mere modifications of the mind? Much as in *CPR*, Kant argues in *Convolut* 7 that determined phenomenal objects are impossible without the synthetic activity of the subject. Likewise, however, the determined phenomenal subject itself is impossible without this activity. Kant says much more about the latter in his *Selbstsetzungslehre*, or doctrine of self-positing, in *Convolut* 7.

In the Transcendental Deduction, Kant discusses two ways in which consciousness can be unified: the synthetic unity of apperception and the analytic unity of apperception. As mentioned in the last section, Kant claims the latter depends upon the former.[109] When it comes to the spontaneity of the subject, in *Convolut* 7 Kant thinks that it involves two acts: (1) an analytic act via concepts by which apperception *thinks* itself as an object, and (2) a synthetic act via intuition by which the subject *cognizes* itself as an object of sense.[110] As Kant says:

> I am conscious of myself (*apperceptio*). I think, i.e. I am to myself an object of the understanding. However, I am also to myself an object of the *senses* and of empirical intuition (*apprehensio*). The thinkable I (*cogitabile*) posits itself as the sensible (*dabile*) and this *a priori* in space and time which are given *a priori* in intuition and are merely forms of appearance.
>
> (*OP* 22:119.10–15)[111]

Kant seems to be reversing his position in *Convolut* 7. Instead of the synthetic unity of apperception (*dabile*) making possible the analytic unity of apperception (*cogitable*), it seems as if the analytic unity is making possible the synthetic unity. Passages like the one above might also provide support for Adickes' view, which holds that the noumenal subject posits the empirical subject *a priori*. If we associate the analytic unity of apperception (or "I think") with the noumenal subject and the synthetic unity of apperception with the phenomenal subject, then it would seem as if the intellectual act of the former is what makes possible cognition of the latter as an object in space and time (i.e., as embodied cognitive subject). Much like Adickes, So-In Choi also believes that the self as object is produced by the subject *a priori*.[112] On the one hand, there is the analytic unity of apperception which involves a logical act of the subject whereby the subject *thinks* of itself according to the categories in a *primitive* manifold of space and time. On the other hand, there is the synthetic unity of apperception which involves a metaphysical act of the subject whereby the subject *cognizes* itself as an object according to the categories in a *derivative* manifold of spatiotemporal appearances.[113] This metaphysical act presupposes, however, the logical act and consequently the synthetic unity of apperception presupposes the analytic unity of apperception. Although there are many similarities between

Choi's account and Adickes' account, Choi does not endorse double-affection.[114] Like Adickes, however, there seems to be no place for genuine *receptivity* in Choi's account. Since the manifold of appearances is derived from the primitive manifold of space and time, it is not simply that the subject provides the *form* of the manifold (space and time), but also the *content* of the manifold itself (appearances). As Choi says, 'The object of the senses or the manifold of the same is now no more from the outside—from affection on the part of the thing-in-itself—given, but is rather from the subject itself posited and given.'[115]

Although this view might sometimes be suggested by the text, its thoroughgoing subjectivism implies a form of absolute idealism contrary to Kant's project in both the Critical and post-Critical periods. Such an interpretation places Kant at odds with himself, which violates my second principle of interpretation (see §0.1). Much of what Kant has to say in the sections leading up to *Convolut* 7 has to do with deducing a transcendental *material* condition of experience and showing how this transcendental material condition might be incorporated with the other transcendental formal conditions of experience. Why would Kant suddenly abandon this material condition of experience in favor of a radical form of idealism? Ultimately, I do not think Kant adopts a form of absolute idealism in *Convolut* 7 and the moving forces of matter continue to play an important role in *Convolut* 7 as the transcendental material condition of experience, just as they have in the sections leading up to *Convolut* 7. Receptivity does not disappear in *Convolut* 7. Quite to the contrary, Kant says repeatedly in *Convolut* 7 that the receptivity of the subject to be affected from without is equally important to the spontaneity of the subject to affect itself from within.[116]

In contrast to the above thinkers, Förster's analysis of the *Selbstsetzungslehre* neither appeals to a noumenal process of affection nor does it ignore the receptivity of the subject to be affected from without. Kant's project in the *Selbstsetzungslehre,* according to Förster, consists of five steps and its aim is to show 'how the I as mere object of thought (*cogitable*) can become an empirical object given in space and time (*dabile*).'[117] Förster believes that Kant continues to hold post-Critically that the analytic unity of consciousness or "I think" is only possible on the assumption of some synthetic unity. As he says:

> The act "I think" could not take place if nothing were given in thought for its synthesis. This first act thus also implies the determin*ability* of myself as object, and the possibility to go beyond what Kant calls a merely logical self-consciousness.[118]

Although the first step of the *Selbstsetzungslehre* is this act of logical self-consciousness, it implies a second step whereby the subject first determines itself as an object by *positing* itself in the pure intuition of space and time. This synthetic unity of consciousness consequently *amplifies* the analytic unity of consciousness beyond the mere concept of a subject. Without the content that this amplification provides, however, there would be nothing for the "I think" to think.

Since the subject is not yet posited as a determined *empirical* object (embodied in space and time), more steps are still needed. The third step of the *Selbstsetzungslehre* requires that the subject *assume* that space is filled with moving forces since it is represented empirically as an object of sense or 'totality of empirically identifiable locations.'[119] It is not merely the bare assumption of the ether that makes the whole of experience possible, however, but the assumption of this ether with certain *attributes*. These attributes are captured by the categories, and the fourth step of the *Selbstsetzungslehre* holds that these categorically determined concepts of the ether must be *inserted* into the undetermined empirical manifold so as to determine to what extent the categories are exemplified in experience.[120] The subject can represent itself as affected from without, however, only insofar as it exercises its own moving forces in reaction to the moving forces which affect it. In this way, the subject posits itself as empirical object (embodied cognitive subject) and cognizes other empirical objects to which it is related in space and time. This is the fifth and final step of the *Selbsetzungslehre.*[121]

Although there is much I agree with in Förster's reconstruction of the *Selbstsetzungslehre,* there are also some important points of difference. When it comes to the former, Förster does an excellent job of explaining the activity on the side of the subject required both for the self-positing of the subject as an empirical object or embodied cognitive subject (determined phenomenal subject) as well as for cognition of empirical objects (determined phenomenal objects) outside of us.

At the same time, the role he assigns to the ether as a mere *assumption* or ideal of reason does not seem sufficient to insure genuine receptivity. Although he does talk about the receptivity of the subject to be affected from without by the moving forces of matter, a commentator like Choi could hold that the ether, as an ideal of reason, is simply posited by the subject. Adickes would hold that it is a mere representation of the noumenal subject which serves as a thing-in-itself for the empirical subject. In either case, the status of the ether is decidedly subjective.

Lord has recently argued that the ether is not an "ideal of reason" in Kant's sense, but rather an "idea of the understanding" in Salomon Maimon's sense.[122] Specifically, it is an idea that makes possible the content of intuition not only transcendentally but genetically. In other words, the ether is not only a condition for the possibility of experience but itself partially produces the objects of experience. In this respect, I certainly agree with Lord. According to Lord, however, the ether is the product of the subject's activity.[123] Whereas the subject is first affected by moving forces from without, through self-affection the subject perceives the effects of the ether which are the subject's own product. In this respect, I have to disagree with Lord's interpretation of *OP.* I see no reason to distinguish the moving forces that originally affect the subject from the ether itself.[124] Although the subject has an *a priori* concept of the ether, the extension of this concept (the ether as transcendental material condition) is not itself the product of the subject's activity. As I have argued, the subject does not produce the ether (or Substance) through its own activity but rather substances (including itself as an embodied cognitive subject). Although actual, the ether is not an object of experience. It is only through the synthetic activities of the subject that the subject perceives determined objects in space and time. As I discussed in §2.2, Kant believes that the ether (or Substance) can only be negatively characterized in itself (imponderable, incoercible, incohesible, and inexhaustible) with respect to the properties that phenomenal objects (or substances) possess, though it makes the positive determination of phenomenal objects possible (i.e., as ponderable, coercible, cohesible, and exhaustible).

Returning to Förster's account, he holds that the force that the subject exercises is primarily *physical*. In what follows, however, I will highlight the *mental* force of apperception involved in self-positing. Finally, Förster's analysis does not emphasize the intrinsic relationship between subjects and objects, and whatever relationship does obtain is weighted heavily toward the subject and its acts. The latter is also a worry for my account of the intrinsic relationship between determined phenomenal subjects and objects in the Critical period. Although I have already argued that the relationship between subjects and objects is *intrinsic*, in what follows I will argue that this relationship is, at its most fundamental level, *symmetrical* as well.

Just as in earlier sections of *OP*, in *Convolut* 7 Kant continues to talk about the ether generating objects and making space sensible.[125] Kant talks not only about self-affection but also about how the subject is affected by objects external to the subject, saying, 'World is the *whole* of sense-objects—thus also including the forces acting on the senses' (*OP* 22:49). Kant defines space and time as *forms* of the subject's *receptivity* to be affected by objects from without.[126] At the same time, he continues to believe the ether is necessary for spatiotemporal unity, saying, 'So as there is only one space and one time—Ether' (*OP* 22:129.7).[127] He also mentions the role of the ether in the thoroughgoing determination of objects.[128] When Kant talks about the moving forces of the ether and their role in affecting receptive subjects from without, I take Kant to be affirming the *actuality* of the ether as a transcendental *material* condition of experience. As he says in *Convolut* 7:

> On this [principle of thoroughgoing determination], and on the principle of the possibility of experience, is founded the idea of the existence of a universally distributed, all-penetrating etc. material which forms the basis of the possibility of there being *one single* experience, and whose existence can thus be comprehended *a priori*.
>
> (*OP* 22:89)

Kant also emphasizes the relational aspect of appearances in *Convolut* 7, saying, 'The objects of representation of intuition are not apprehensible objects outside this object, but the relations of objects to the subject—not as things in themselves = X but as appearances' (*OP* 22:417). Much as in *CPR*, he reiterates the idea that 'the understanding is here occupied with mere relations' (*OP* 22:97–98). Although we already know that Kant is committed, in the *Selbstsetzungslehre,* to the idea that subjects make determined phenomenal objects possible, he also holds that the subject must be affected from without in order to posit itself as a determined phenomenal subject (embodied cognitive subject).[129] As in the Critical period, appearances are causal and spatiotemporal intrinsic relations.

Although appearances have the same basic character in the post-Critical period, Kant does introduce the distinction between direct and indirect appearances in *Convoluts* 10–11 (see §4.1). Determined phenomenal objects are indirect appearances (or "appearance of an appearance") produced by the physical activity of the ether (direct appearance) in conjunction with the subject's mental activity. Although Kant only fleetingly mentions the distinction between direct and indirect appearances in *Convolut* 7, it still seems to be operating in the background of Kant's discussion of self-positing.[130] While *Convoluts* 10–11 describe the post-Critical process that produces determined phenomenal *objects,* the *Selbstsetzungslehre* of *Convolut* 7 adds a description of the process that produces determined phenomenal *subjects.* We already know from *Convolut* 10, however, that the determined phenomenal subject (subject made into an object for itself) is an appearance of an appearance much as the determined phenomenal object:

> The appearance of appearances (that is, how the subject is mediately affected) is metaphysically [the same] as how the subject makes itself into an object (is conscious of itself as determinable in intuition).
>
> (*OP* 22:326)[131]

Given Kant's description of the ether in the *Elementarsystem,* the *a priori* proof strategy of the Ether Deduction, and the role of the ether in thoroughgoing determination (without itself being subject to the principle of thoroughgoing determination), one should consider the ether (or Substance) as an actual though undetermined phenomenal object. The moving forces of the ether stand in direct spatiotemporal causal-dynamic appearance relations with an undetermined phenomenal subject. It is the ether which is the source of the representations (or subjective perceptions) that apperception organizes. Determined phenomenal objects are the result of this organization and emerge from the ether. In this sense, the ether is both a transcendental condition for and a genetic source of determined phenomenal objects.[132]

This is not to say, however, that activity of the ether is sufficient for the generation of determined phenomenal objects. In accordance with Kant's discussion in *Convoluts* 10–11 (see §4.3), through the mental activity of apperception, the subject (1) inserts the categories (in their role as Principles) into subjective perceptions generating a distributive unity of objective perceptions, and (2) inserts the *a priori* concept of the ether into the distributive unity of objective perceptions. This produces both determined phenomenal objects (substances) as well as a determined phenomenal subject (embodied cognitive subject), which are intrinsically related through indirect spatiotemporal causal-mechanical appearance relations in a single spatiotemporal framework (*pace* Kant's Critical view).

This is the sense in which 'we make everything ourselves.' We make everything in the sense of organizing the subjective perceptions that the ether provides, but we do not make these subjective perceptions. It is only through the physical activity of the ether (affection) and the mental activity of the subject (self-affection), that the subject is able to *experience* both determined phenomenal objects as well as itself as a determined phenomenal subject in a unified spatiotemporal framework.

Although I have explained at some length, in the earlier sections of this chapter, how determined phenomenal objects and subjects are intrinsically related, I believe that we should also consider the undetermined phenomenal object and subject as intrinsically related. Without the moving forces of the ether affecting the undetermined phenomenal subject, the subject would have *nothing* to think about, i.e., the mental activity of apperception would have no content and thought would be empty. Kant's post-Critical theory of affection allows us to view his description of affection, at the beginning of the B-edition of *CPR,* in a new light. Within the post-Critical view, it is the moving forces of the ether that:

> Stimulate our senses and in part themselves produce representations, in part bring the activity of our understanding into motion to compare these, to connect or separate them, and thus to work up the raw material of sensible impressions into a cognition of objects that is called experience. (*CPR* B1)

In other words, without the moving forces of the ether, the activity of apperception would never be brought into motion. As I have suggested above but will emphasize below, without content to operate over, apperception is the subjective correlate to the transcendental object = X. Although this makes clear how the mental activity of the subject depends on the physical activity of the ether, how does the ether depend on the subject?

One way in which the ether depends on the subject can be gleaned from its empirical reality (§4.1). Just as in the Critical period, Kant continues to maintain post-Critically that space and time are transcendentally ideal (i.e., the subject's forms of intuition). Without the necessary possibility of subjects like us, there would be no space and time for the ether to realize empirically. Put slightly differently, without the necessary possibility of subjects like us, the ether would be neither omnipresent (spatial) nor sempiternal (temporal). These are, however, essential properties of the post-Critical ether, properties that inextricably tie it to Kant's Critical-era conception of Substance.

The modal status of the ether can also be viewed as a way in which it depends upon the subject. The ether is materially necessary and a transcendental condition for experience (see §3.4). The material transcendental necessity of the ether can be understood in the following way: in all and only possible worlds where there is a unity of experience and creatures cognitively similar enough to us, the ether exists. The existence of the ether is only proven relative to the conditions of possible experience for subjects like us and cannot be affirmed outside of this relationship. As the source of direct appearances, the ether is actual given its connection with 'the material conditions of experience (of sensation)' (A218/B266).[133]

Both the ether and apperception are undetermined without one another. Just as apperception can only be positively determined into an embodied cognitive subject through the physical activity of the ether, so too can the ether only be positively determined into physical bodies through the mental activity of apperception. Both physical bodies and embodied cognitive subjects emerge from the joint activity of the ether and apperception and it is only through this joint activity that anything determinate can emerge from them. A bit earlier, in *Convolut* 11, Kant makes clear that he views these different forms of activity as being reciprocal and simultaneous:

> Positing and perception, spontaneity and receptivity, the objective and subjective relation, are simultaneous; because they are identical as to time, as appearances of how the subject is *affected*—thus are given in the same *actus* and are in progression toward experience (as a system of perceptions).
> (*OP* 22:466)[134]

Consequently, just as determined phenomenal objects and subjects are intrinsically related through indirect appearances, so too are undetermined phenomenal objects and subjects intrinsically related through direct appearances. Whereas I represented the relationship between the ether or substances, on the one hand, and subjects, on the other, as a *one-way* relation in the *Convoluts* 10–11 (see the diagram at the end of §4.3), it should really be viewed as a *two-way* relation at two different levels:

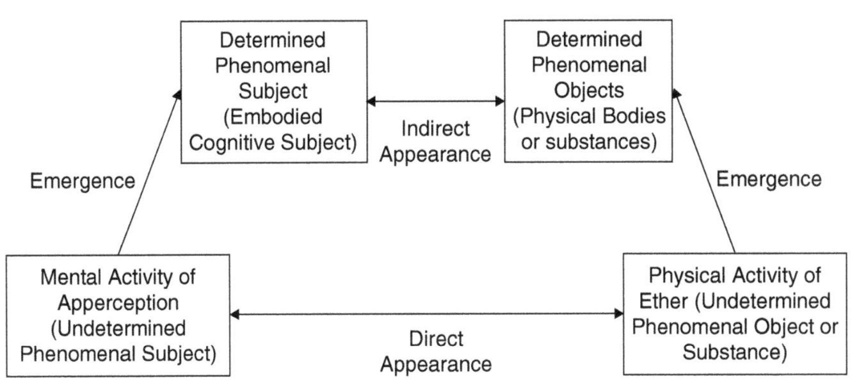

Most importantly, one should note that Kant's post-Critical view of affection requires appeal to nothing outside of the phenomenal world. Contrary to the double-affection view, the original act of apperception ("I think") does not entail that there is some transcendent subject-in-itself or positively noumenal substance that acts as this originator ("I am"). In *Convolut* 7, Kant returns to the criticisms of Descartes that he first made in the Paralogisms section of *CPR*. In the latter, Kant claims that Descartes' inference from the *subject* of the judgment "I think" to a *substance* corresponding to this subject "I am" is fallacious.[135] According to Kant, the "I think" involves only the relations of concepts in a judgment and does not bring with it anything from experience. Existential claims (e.g., "I am"), however, must always involve the unification of concepts and sensible intuitions. He repeats this claim in *Convolut* 7, saying that there is no valid logical inference from the "I think" to the "I am."[136] If the latter is meant substantively, it requires intuition of oneself as an object in space and time. As Kant says:

> The logical act, I *think* (*apperception*) is a judgment (*iudicium*), but not yet a proposition (*propositio*), not yet an act of the faculty of knowledge (*facultas cognoscendi*) through which an object is given; rather, it is only thought in general. It is, according to its form, a logical act, without content (*cogitans sum, me ipsum nondum cognosco*), even less is it a rational inference: I think therefore I am (*ratiocinium*). I, the subject, makes itself into an object according to the rule of identity. Two elements belong to knowledge (*cognitio*), intuition and concept, a representation through which an object is given and another by which it is thought.
>
> (*OP* 22:95)

Since apperception is a function of the understanding prior to sensibility, Kant cannot characterize it as an *object,* and instead characterizes it simply in terms of its *activity.* Kant talks about the mental activity of the subject as something the subject produces from itself prior to positing itself as an object (i.e., as an embodied cognitive subject). As he says, 'One does not begin from objects, but rather from the system of the possibility of constituting one's own thinking subject, and one is oneself the originator of one's power of thought' (*OP* 22:79). In this respect, I do not think one should characterize mental activity itself as something that emerges from the ether.[137] Although determined phenomenal objects and subjects emerge from the physical activity of the ether in conjunction with the mental activity of apperception, the latter form of activity does not emerge from the ether. The autonomy of apperception is important for insuring both the genuine spontaneity of subjects to affect themselves from *within* as well as their genuine receptivity to be affected from *without.* Even though the moving forces of matter can unite both inorganic bodies and organic bodies in one system of nature (see §2.3 and §4.4), they cannot explain the mental powers of the subject.[138] Put differently, although the subject insofar as it is *embodied* (an organism) emerges from the moving forces of the ether, the embodied subject insofar as it is *cognitive* does not.[139] Parallel to the relationship between determined phenomenal objects (or substances) and the activity of the ether (Substance), the relationship between the determined phenomenal subject and the activity of apperception should be considered a case of emergence (rather than reduction or supervenience). The determined phenomenal subject possesses properties that apperception alone does not possess, e.g., in the Refutation of Idealism, consciousness of its existence as determined in time (contra reduction).[140] Likewise, as an *embodied* subject, it possesses its own causal efficacy in the natural world (contra supervenience). In either the case of determined phenomenal objects or determined phenomenal subjects, however, they emerge from the *joint* activity of the ether and apperception.

The activity of apperception is expressed through acts of synthesis, according to concepts, whereby the subject unifies the manifold of intuition and posits phenomenal objects including itself as object. It is only through the unification of the given manifold of intuition that the subject can posit itself as an object. Insofar as the subject is spontaneous, Kant believes the subject is a thing-in-itself understood as an 'act of the understanding = X.' Given what has been said above, however, the subject-in-itself should not be considered a positive noumenon, but rather only a transcendental object or negative noumenon. Just as the transcendental object is a placeholder for the object of representation, so too is the subject prior to self-positing a mere transcendental placeholder for the activity of apperception. Whether we are talking about the activity of apperception or the activity of the ether, internal or external affection, considered on their own 'the affecting object is = X' (*OP* 22:36). We can only cognize ourselves as subjects insofar as we appear to ourselves through the combination of internal (self-positing) and external (locomotive substances or dynamical ether) affection. The activity of apperception, however, is nothing in itself, but only a conceptual placeholder. As Kant says in *Convolut* 7:

> [The subject's] consciousness of itself (*apperceptio*), insofar as it is affected, is the representation of the object in appearance. However, insofar as it is the subject which affects itself, it is equally to be regarded as the object in itself = X.
>
> (*OP* 22:78)[141]

Although apperception and the ether are transcendental conditions for the possibility of the phenomenal world, this does not entail that they are independently existing things-in-themselves or *positive* noumena since, in themselves, neither one is a determinate thing. As an extension of Kant's argument in the Paralogisms, one might say that any inference from the activity of the ether or of apperception to some transcendentally real object that underlies this activity would be fallacious. As mentioned above, Kant's post-Critical ontology is ultimately one of *force* whether physical (the moving forces of the ether) or mental (the spontaneity of apperception). Considered in themselves and apart from their relationship to one another, however, they are only *logical* (conceptual) grounds (= X). They become *real* grounds only through this relationship, i.e., insofar as they mutually determine one another and so jointly produce the phenomenal world through their activity.

5.6 CONCLUSIONS ABOUT VAIHINGER'S TRILEMMA

Although I believe early commentators on *OP* were right to look to it for a solution to the problem of affection, I believe the solution they offered, double-affection, is both philosophically problematic as well as textually unsupported. Although the theory of affection that I have developed involves two-levels of affection (direct and indirect appearance), both levels of affection occur wholly within the confines of the phenomenal world.[142]

Returning one last time to Vaihinger's trilemma, the thing-in-itself cannot be understood as the affecting object, insofar as the thing-in-itself is to be understood as a *positive* noumenon equipped with intrinsic non-relational properties. Although Langton admits that subjects cannot cognize *what* these intrinsic properties are, she still goes too far in claiming that subjects can cognize *that* there are things-in-themselves with intrinsic properties and *that* these things-in-themselves affect receptive subjects in sensibility. As Allison argues, the cognition that there are things-in-themselves with intrinsic properties would require an illegitimate application of the unschematized category of substance. And as I have argued, assuming that affection is a causal relation, the cognition that these things-in-themselves affect us in sensibility would require an illegitimate application of the unschematized category of causation. Consequently, there seems little way to make sense of affection via the first horn.

Although I argued that Langton is *minimally* committed to defending the first horn of Vaihinger's trilemma, upon reflection she may in fact be committed to defending a version of the third horn. For Langton, appearances are a special subclass of phenomena where the latter are constituted by the relations between things-in-themselves. These things-in-themselves are, in turn, substances equipped with intrinsic non-relational properties. In Kant's language, they are *positive* noumena. Consequently, it seems an appearance must also be understood as a relation between positive noumena. Insofar as a human mind is one of the relata in an appearance relation, the human mind must be understood as a positive noumenon, a substance equipped with intrinsic non-relational properties. The only candidate for such a thing under Kant's view would be the noumenal subject. Appearances are then relations between things-in-themselves (noumenal objects) and noumenal subjects which give rise to the experience of phenomenal objects. This is quite close to the double-affection view. Despite the virtues of double-affection, the third horn of Vaihinger's trilemma seems doomed to failure for the reasons mentioned above (see §4.1).

My solution to the problem of affection shares much in common with the second horn insofar as I affirm that affection takes place wholly within the phenomenal world. It attempts to avoid the problems associated with the second horn, however, by understanding appearances not as *objects* themselves, which could lead to Berkeleyan phenomenalism, but rather as spatiotemporal causal *intrinsic* relations, between phenomenal objects and subjects. At the same time, my solution avoids the first horn since the affecting object is not an independently existing thing-in-itself or *positive* noumenon. Even so, the thing-in-itself still has a role to play as a *negative* noumenon or transcendental placeholder for the object of representation as well as for the subject of representation. The intrinsic nature of the relationship between phenomenal objects and subjects comes out very clearly, I believe, in the Transcendental Deduction of the Categories. Without objects of representation, there would be no determined subject of representation. Without the subject of representation, there would be no determined objects of representation.

In *OP*, Kant expands this theory by explaining how phenomenal objects and subjects are intrinsically related to one another at two-levels. The ether is an undetermined phenomenal object which stands in *direct* appearance relations with the undetermined phenomenal subject. The physical activity of the ether produces the representations (or subjective perceptions) that the mental activity of apperception organizes into determined phenomenal objects which stand in *indirect* appearance relations (objective perceptions) with the determined phenomenal subject that apperception posits.

Just as determined phenomenal objects and determined phenomenal subjects are intrinsically related through indirect appearances (causal-mechanical), so too are the ether and apperception intrinsically related through direct appearances (causal-dynamical). Although I already explained the former relationship above (§5.4), one can see the latter relationship in the way that the ether and apperception mutually determine one another. Both the ether and apperception are undetermined in themselves, but mutually determine one another through their activity in the phenomenal world. Without the activity of the ether, a determined phenomenal subject (embodied cognitive subject) could not emerge from the activity of apperception. Without the activity of apperception, determined phenomenal objects (substances) could not emerge from the activity of the ether. To borrow a line from Billie Holiday, when it comes to appearances (either direct or indirect), it is all or nothing at all.

NOTES

1. F.H. Jacobi, *Werke* (Darmstadt: Wissenschaftliche Buchgesellschaft, 1968), vol. 2, 304.
2. Vaihinger, *Commentar zu Kants Kritik,* vol. 2, 53.
3. Kant discusses the sensible conditions for the application of the categories at length in the Schematism section of *CPR*. See *CPR* A137–147/B176–187. See

also *CPR* A246/B303 where Kant claims that the categories are only of empirical use.

4. Van Cleve, *Problems from Kant,* 163.
5. Commentators break down into two main camps. The first defends a "one-world" interpretation. Some commentators, like Harold Langsam, adopt a one-world *ontological* reading of Kant whereby things-in-themselves are the only objects that exist. Things-in-themselves are capable of affecting cognitive subjects since the latter apply the *unschematized* (temporally undetermined) category of causation to the former. See Harold Langsam, "Kant, Hume, and Our Ordinary Concept of Causation," *Philosophy and Phenomenological Research* 54, no. 3 (1994): 625–647. I will discuss this use of the categories in connection with Rae Langton's own one-world view below. Other commentators, like Allison, however, adopt a one-world *epistemological* reading whereby one can *consider* the same object from two different standpoints or aspects. See Allison, *Kant's Transcendental Idealism,* 66–70. The second camp defends a "two-world" interpretation of Kant whereby appearances and things-in-themselves are somehow distinct from one another. Proponents of the double-affection view, like Adickes, are typically understood as defending this position at an ontological level. See Adickes, *Kants Lehre von der doppelten Affektion.* As I will argue below, however, Adickes seems to incorporate something like Allison's dual-aspect theory within his theory of double-affection and this might suggest that Adickes actually holds a one-world view. More recent commentators like Desmond Hogan and Van Cleve argue for varieties of the two-world view. Hogan argues for noumenal affection of the phenomenal subject through the free acts of the noumenal subject. Although he does not talk about the relationship between noumenal objects and noumenal subjects, the relationship between noumenal subjects and phenomenal subjects reflects the double-affection view. See Desmond Hogan, "Noumenal Affection," *Philosophical Review* 118, no. 4 (2009): 501–532. For Van Cleve, phenomenal objects are virtual or intentional objects that are the result of noumenal construction. At the same time, however, the two-world distinction is semantic and not ontological for Van Cleve. See Van Cleve, *Problems from Kant,* 120 and 150–156. Vaihinger seems to have been the first, however, to consider Kant's view of phenomenal objects as similar to Brentano's view of intentional objects. See Vaihinger, *Commentar zu Kants Kritik,* 6, 34, 56. What is most important to note, however, that all of the above commentators, with the exception of Langton, understand appearances as *particulars* and not as *relations.*
6. It is important to distinguish between the different senses of "property" and "relation" that I will be using in this chapter. Monadic or non-relational properties are one-place properties (e.g., X is simple). When intrinsic, these properties are essential to the objects that have them. Relational properties are multi-place properties (e.g., X is to the left of Y) which may or may not be essential to the objects so related. Extrinsic relations are inessential to the objects so related. Intrinsic relations are, however, essential to the objects so related. What this means, for my purposes, is that not all relations are extrinsic, nor are all intrinsic properties non-relational.
7. Two points on terminology: 1) By characterizing the undetermined phenomenal subject primarily in terms of its activity (apperception), I do not mean to imply that this subject is not receptive to affection in outer sense. Being affected by the moving forces of matter in outer sense is necessary for the subject to determine itself as an object in outer sense and be aware of its identity over time in inner sense. 2) Although the undetermined phenomenal object (ether) possesses categorical properties, these are properties that objects of experience

(substances) cannot possess (see §2.2). The ether is not itself an object of experience. This is important since Kant claims, in the A-edition Paralogisms, that 'I cannot cognize as an object itself that which I must presuppose in order to cognize an object at all' (*CPR* A402). Since Edwards mistakenly interprets Kant as holding that the ether is an object of experience, he claims that the Ether Deduction runs afoul of this passage. See Edwards, "Spinozism, Freedom, and Transcendental Dynamics," 73n39.

8. Although Kant himself does not use these exact words to describe positive noumena, he does say that the transcendent ideas of the Antinomies (e.g., God and the noumenal subject) have merely intelligible objects which are things determinable by their 'distinguishing and inner predicates' (*CPR* A565–566/B593–594). We will return to the idea of positive noumena or *substantiae noumena* again below (§5.4).
9. See *CPR* A533–534/B561–562. For an excellent discussion of the many contexts in which Kant discusses things-in-themselves, see Robert Adams, "Things in Themselves," *Philosophy and Phenomenological Research* 57, no. 4 (1997): 801–825. Hogan argues for noumenal affection on practical grounds within the context of the Critical philosophy. See Hogan, "Noumenal Affection."
10. See Adickes, *Kants Lehre von der doppelten Affection,* 27. Kant distinguishes between the terms "transcendent" and "transcendental" in *CPR*. A *transcendent* principle can find application only *beyond* the bounds of sense. A *transcendental* principle, however, is properly applied only *within* the bounds of sense. See *CPR* A295–296/B352–353.
11. Adickes, *Kants Lehre von der doppelten Affection,* 47 and 77–80.
12. Adickes, *Kants Lehre von der doppelten Affection,* 79.
13. Adickes comes close to endorsing the position that I attribute to Kant insofar as he recognizes that the "appearance in itself" and the "empirical self" reciprocally affect one another. At the same time, he believes that empirical affection, on its own, absent affection through things-in-themselves, is insufficient to avoid Berkeleyan idealism. See Adickes, *Kants Lehre von der doppelten Affection,* 45–46.
14. Adickes, *Kants Lehre von der doppelten Affection,* 65–70.
15. For appearances as affecting objects see, for example, *CPR* A28 and A213/B260. For things-in-themselves as affecting objects see, for example, *CPR* A190/B235 and A288/B344.
16. Kemp Smith, *A Commentary to Kant's Critique,* 612–613. See also Adickes, *Kants Opus postumum,* 18.
17. Adickes, *Kants Lehre von der doppelten Affection,* 32. See also Adickes, *Kants Opus postumum,* 851. He admits, however, that the theory of double-affection, as he understands it, is never fully developed in *OP.* See Adickes, *Kants Opus postumum,* 419.
18. Adickes believes these logical-teleological relations are also necessary to insure the necessity of empirical law and to guarantee the affinity of the empirical manifold. See Adickes, *Kants Lehre von der doppelten Affection,* 78, 84, 89–94. More recently, Westphal has made a similar point when it comes to insuring the affinity of the empirical manifold. See Westphal, *Kant's Transcendental Proof of Realism,* 110–116. As I argued in the last chapter, the ether serves as the ultimate ground of objectivity in the post-Critical period.
19. Adickes, *Kants Lehre von der doppelten Affection,* 8–9.
20. For example, see *OP* 22:319–320, 22:326, 22:367–368.
21. Translation is mine. Quoted by Adickes, *Kants Opus postumum,* 654. Kant's definition of the thing-in-itself in *Convolut* 7 is very similar to his definition of

the transcendental object in *CPR*. We will return to relationship between the thing-in-itself and the transcendental object below (§5.4).

22. As we saw in the Introduction (§0.2), this created tension between Buchenau and Adickes that contributed to Adickes' decision to resign from the project of editing *OP*.
23. See Adickes, *Kants Opus postumum,* 702–718.
24. Quoted by Adickes, *Kants Opus postumum,* 648.
25. Kemp Smith, *A Commentary to Kant's Critique,* 618.
26. Kemp Smith, *A Commentary to Kant's Critique,* 636.
27. Adickes, *Kants Lehre von der doppelten Affection,* 92.
28. Robert Paul Wolff, *Kant's Theory of Mental Activity* (Cambridge, MA: Harvard University Press, 1963), 173. Kemp Smith also acknowledges that double-affection depends on this rather tenuous analogy. See Kemp Smith, *A Commentary to Kant's Critique,* 278.
29. Allison also poses this objection to double affection in *Kant's Transcendental Idealism,* 66.
30. See Vaihinger, *Kommentar zu Kants Kritik,* vol. 2, 51–53.
31. See Gerold Prauß, *Kant und das Problem der Dinge an sich* (Bonn: Bouvier, 1974). Allison acknowledged his debt to Prauß. See Allison, *Kant's Transcendental Idealism,* 50–52.
32. Adickes, *Kants Lehre von der doppelten Affection,* 3. Translation is mine.
33. For the purposes of the analogy, the focus should be placed not on the contrast between the *portrait* of the object and the *object* being portrayed. The portrait of something is clearly a different object than whatever it is that the portrait portrays (e.g., a portrait of Kant hanging on the wall is a different object from Kant). The focus should rather be placed on the contrast between the portrait of the *object* and the *object* being portrayed. For Adickes, the object in the portrait (thing as it appears) is the same as the object being portrayed (thing-in-itself), though the former requires the artist's (subject-in-itself) interpretation of the latter.
34. Adickes, *Kants Lehre von der doppelten Affection,* 51.
35. Nicholas Stang also notes that Adickes subscribes to a metaphysical dual-aspect view. See Nicholas Stang, "Adickes on Double Affection," in *Kant und die Philosophie in weltbürgerlicher Absicht: Akten des XI. Internationalen Kant-Kongresses,* ed. Stefano Bacin and others (Berlin: De Gruyter, 2013), vol. 2, 787–798.
36. Adickes, *Kants Lehre von der doppelten Affection,* 2 and 28.
37. See *CPR* Bxxvi–xxvii.
38. Paul Abela, *Kant's Empirical Realism* (Oxford: Oxford University Press, 2002), 37–40.
39. Van Cleve, *Problems from Kant,* 8.
40. Hanna, *Kant and the Foundations of Analytic Philosophy* (Oxford: Oxford University Press, 2001), 109.
41. See *CPR* Bxxvi–xxvii. As I will argue below (§5.4), the relationship between the thing-in-itself and the subject in this passage should be viewed as logical and not causal. Although this might remind one of the logical-teleological relations that are supposed to obtain between positively noumenal objects and subjects under Adickes' view, the relationship that I describe will not require positively assertoric commitments beyond the phenomenal world.
42. Rae Langton, *Kantian Humility: Our Ignorance of Things in Themselves* (Oxford: Oxford University Press, 1998), 19.
43. See *CPR* A274/B330.
44. Langton, *Kantian Humility,* 12.
45. Langton spends some time discussing two arguments against the reducibility of relations. Although she admits that there are problems with both arguments,

she adopts the version that rejects the bilateral reducibility of relations where the latter are understood as causal powers (i.e., causal powers that are irreducible to the intrinsic properties of the relata of causal relations), given the fact that God could have chosen natural laws that would not allow for causal powers even if the intrinsic properties of the relata were held fixed. See Langton, *Kantian Humility,* 112–123.

46. Langton, *Kantian Humility,* 19.
47. Langton, *Kantian Humility,* 23.
48. Langton, *Kantian Humility,* 13.
49. See Langton, *Kantian Humility,* 48–52.
50. See Allison, *Kant's Transcendental Idealism,* 17–18. Critics have posed a number of objections to Langton's humility thesis. For example, and in contrast to Allison, Van Cleve believes that Langton's thesis is far too humbling, arguing that it precludes knowledge even of our own mental content. See Van Cleve, "Receptivity and Our Knowledge of Intrinsic Properties," *Philosophy and Phenomenological Research* 65, no. 1 (2002): 232.
51. Langton, *Kantian Humility,* 13.
52. See *MFNS* 4:556. What is required is that both relata are locomotive. Although this remains true, post-Critically, with regard to *indirect* appearances, Kant allows for *direct* appearances of the ether notwithstanding the fact that the latter is not locomotive. Unlike Langton's thing-in-itself, however, the ether is spatiotemporal.
53. Langton, *Kantian Humility,* 7.
54. Langton, *Kantian Humility,* 13.
55. For the logical table of judgments, see *CPR* A70/B95.
56. I believe this also rules out Langsam's one-world ontological interpretation mentioned above.
57. See also A251–252, A265/B321, B306–308, and A284–85/B340–41.
58. See Langton, *Kantian Humility,* 186–204.
59. See also A44–A45/B61–B62.
60. See *CPR* A264/B320.
61. It is important to note that outer relations need not be *extrinsic* relations. The concept of an extrinsic relation has nothing to do with spatial character. As I will argue below, these outer relations are, in fact, *intrinsic* relations.
62. For Kant's explicit discussion of "*substantiae noumena*" see *CPR* A276/B332.
63. See *MFNS* 4:544. See also Kant's definition of the 'mere concept of matter,' in *CPR,* as 'impenetrable lifeless extension' (*CPR* A848/B876). Here I am following Friedman, *Kant's Construction of Nature,* 346–347. Although Kant does not suggest that we can cognize the absolutely internal in *OP,* he does abandon the inertness of matter with his dynamic ether theory. The *physical* activity of the ether, however, should not be understood in terms of *mental* activity. I already touched on this issue in §2.3 and will return to it in §5.5.
64. See *CPR* A20/B34.
65. See also *CPR* A267/B323.
66. Hanna describes this position as "weak transcendental idealism." This is the position that space and time can exist in a possible world (including the actual world) even if no cognitive subjects like us existed in that world, provided that if cognitive subjects like us existed in this world they could correctly represent space and time. See Hanna, *Kant, Science, and Human Nature,* 299–312. One might still push the modal issue, however, claiming that, under my view, it makes little sense to talk about actual objects without reference to actual subjects (e.g., the actual world before cognitive subjects) since actual objects must bear intrinsic relations to actual subjects and vice-versa. At this point, I think it is important to note Kant's rather unique way of understanding modality. For Kant, the domains of the necessary, actual, and possible are *coextensive* and the

difference between these modalities is simply *intensional*. The modal categories have different meanings insofar as they describe different cognitive relations or ways of considering objects. Consequently, regardless of whether you are talking about the necessity, actuality, or possibility of a given object, you are always presupposing some *relation* to the subject and, for Kant, it makes little sense to consider the modal status of some object without reference to some subject doing the *considering*. In the above example, Kant would say that one can consider actual objects in the distant past prior to the existence of cognitive subjects via the connection that the former has to perception in accordance with natural law. This is the upshot of Kant's Second Postulate of Empirical Thought describing the category of actuality. For Kant's discussion of the coextensive nature of the modal categories and their intensional differences see *CPR* A230–235/B282–287. See also Sebastian Gardner's helpful discussion of the modal categories in Gardner, *Kant and the Critique of Pure Reason,* 197–198.

67. Langton, *Kantian Humility,* 17–24. One of the passages that Langton repeatedly uses to support the idea that substances must have intrinsic non-relational properties is *CPR* A274/B330 (first quoted by Langton, *Kantian Humility,* 16) from the Amphiboly section. An interpretive difficulty with this section, however, is that it is often hard to disentangle Kant's view from the Leibnizian view he is criticizing. When it comes to the passage that Langton quotes, Graham Bird argues that Langton mistakes the Leibnizian position that Kant rejects for Kant's own position. See Graham Bird, "Review of Kantian Humility," *The Philosophical Quarterly* 50, (2000): 106.
68. For different contexts in which Kant makes use of empirical affection, see *CPR* B208, A213/B260, and A491/B520. One of the most thorough examinations of Kant's views on empirical affection can be found in Adickes, *Kants Lehre von der doppelten Affection,* 5–26.
69. Hanna takes a different approach to this problem, holding that objects and embodied subjects just are complex events of the dynamic ether. See Hanna, *Kant, Science, and Human Nature,* 397–398, 445. Although this is an intriguing solution, as mentioned above (§2.2), I do have reservations concerning Hanna's ontology.
70. For Watkins, a Kantian event is not the state of a substance at a particular time, but rather a change from one temporally determinate state to another. See Watkins, *Kant and the Metaphysics of Causality,* 12 and 236
71. See Watkins, *Kant and the Metaphysics of Causality,* 284–285.
72. Watkins' interpretation of the Analogies also provides a foil to Westphal's interpretation where all three Analogies can be described simply in terms of Substance (see §1.1).
73. Bracketed items are my addition.
74. I thank Reed Winegar for pointing out this problem.
75. See Boyle, "Origin of Forms and Qualities," 23. I thank Walter Ott for pointing out this example.
76. David Hume, *A Treatise of Human Nature,* ed. P.H. Nidditch (Oxford: Oxford University Press, 1978), 636.
77. Berkeley's view on existence can be interpreted in two different ways: (1) to be is to be perceived or to be a perceiver, or (2) to be is to be capable of being perceived or to be capable of perceiving. The former interpretation is suggested by Berkeley's discussion in the *Dialogues* where he claims that all sensible things must be perceived by God in order to exist. See *Three Dialogues between Hylas and Philonous,* from *The Works of George Berkeley Bishop of Cloyne,* eds. A.A. Luce and T.E. Jessop (London: Thomas Nelson and Sons, 1948–1957), vol. 2, 212–213 (II.18). The latter interpretation is suggested by Berkeley's discussion in the *Principles* where he claims that sensible things must only be capable of being

perceived in order to exist (e.g., that the table in his study exists since it would be perceived if he were in his study to perceive it). See *A Treatise Concerning the Principles of Human Knowledge* from *The Works of George Berkeley Bishop of Cloyne,* vol. 2, 42 (Part I, §3). If nothing else, the second interpretation would pose problems for Berkeley's argument for God via the first interpretation. In any case, for present purposes, I assume the first interpretation.

78. See Berkeley, *Principles,* Part I, §25.
79. See Berkeley, *Principles,* Part I, §27–32. According to Berkeley, although we can produce ideas of imagination through our own will, the ideas of sense are produced by the will of God. In either case, however, the ideas themselves are causally inert.
80. Van Cleve, *Problems from Kant,* 170. As Van Cleve interprets Kant, it is possible to abstract the structure (appearances) from the stuffing (things-in-themselves) since the structure is *extrinsic* to the stuffing. On my view, however, the structure (appearances) is *intrinsic* to the stuffing (phenomenal objects and subjects) and so the structure cannot be abstracted from the stuffing.
81. See *CPR* B146–147. As we will see in the next section, the ether can be considered a perpetually undetermined object of representation. When I refer to the interdependency of objects and subjects in this section, however, I am referring to the interdependency of *determined* objects and subjects.
82. See *CPR* B137.
83. Synthesis, in the most general sense, is putting together representations to discover what is common between them. See *CPR* A77/B103.
84. Gardner also argues that the subject and object make one another possible. See Gardner, *Kant and the Critique of Pure Reason,* 157–160. Similarly, Allison argues that there is a reciprocal relationship between the synthetic unity of apperception and the representation of objects. See Allison, *Kant's Transcendental Idealism,* 173–178. As mentioned above, however, Adickes comes closest to articulating the position that I attribute to Kant.
85. See *CPR* B146–147.
86. See *CPR* B275–276. Friedman connects the Refutation to the General Remark to the Principles (*CPR* B291) and to *MFNS* (4:471) insofar as they all likewise emphasize the epistemic priority of outer sense to inner sense. See Friedman, *Kant's Construction of Nature,* 6–7.
87. Although the Refutation might seem to suggest that these phenomenal objects must be spatially distinct from the embodied cognitive subject, Hanna has argued that the conclusion of the Refutation would be satisfied simply by the embodied cognitive subject. See Hanna, *Kant, Science, and Human Nature,* 68–80. In such a case, however, one's own body should be considered the determined phenomenal object required for the determination of the subject's identity in time (determined phenomenal subject). Likewise, as we will see in *Convolut* 7, without the self-positing activity of the subject, the subject would not itself be embodied (a determined phenomenal object).
88. See, for example, *CPR* Bxxvi, A249, A251–252.
89. Bird, "Review of Kantian Humility," 105.
90. See also A190/B235, A280/B336, and A283/B339.
91. For more on Kant's contrast between transcendental reality/empirical ideality (metaphysical realism of Rationalism/phenomenalism of Empiricism) and transcendental ideality/empirical reality (Kant's own Critical view), see *CPR* A367–380.
92. There are several places where Kant seems open to analytic judgments involving the *concept* of a thing-in-itself, e.g., *CPR* B149, A433/B461, and A635/B664. Even so, as Kant makes clear in his criticisms of the ontological argument, such analytic judgments cannot have existential import. See *CPR* A597/

B625 as well as his comments at the end of the Phenomena/Noumena section at *CPR* A258–259/B314–315.

93. Gerd Buchdahl also recognizes the *conceptual* nature of the transcendental object. See Gerd Buchdahl, *Kant and the Dynamics of Reason: Essays on the Structure of Kant's Philosophy* (New York: Blackwell, 1992), 161.
94. See also *CPR* A372.
95. Allison, *Kant's Transcendental Idealism,* 63. Jay Rosenberg characterizes the relationship between the two concepts somewhat differently. Whereas the concept of a negative noumenon results from abstracting away the sensible conditions whereby an object can be given, the concept of the transcendental object results from abstracting away the conceptual conditions whereby an object can be thought. See Jay Rosenberg, *Accessing Kant: A Relaxed Introduction to the Critique of Pure Reason* (Oxford: Oxford University Press, 2005), 252–253.
96. *CPR* A250–252. For other contexts in which Kant uses "transcendental object" and "noumenon" interchangeably, see *CPR* A358 and A288/B344–345. It is important to note that Kant claims in the paragraph immediately following the above quoted passage that the transcendental object *cannot* be called a noumenon. Buchdahl overcomes the seeming contradiction by noting that the distinction between negative and positive noumena is only implicit in this section of the A-edition. Once one has this distinction in mind, however, there is little problem seeing how the transcendental object could be equivalent to the negative noumenon while certainly different from a positive noumenon. See Buchdahl, *Kant and the Dynamics of Reason,* 84–85.
97. In the B-edition, Kant characterizes the negative noumenon as 'a thing insofar as it is not an object of our sensible intuition because we abstract from the manner of our intuition of it.' See *CPR* B307. This might suggest that the object has a way that it is, some transcendental reality, independent of the way in which we intuit it. It is important to note that this is an act of abstraction, however, and so results only in the concept of an object in general considered independently of any particular features in sensibility. This concept need not have a referent, however, in order to function as a concept. In fact, Kant describes the negative noumenon in precisely this way before the passage quoted above. See *CPR* B306.
98. Although I am arguing that this is already Kant's position in the Critical period, Sang-Bong Kim ascribes a similar position to Kant in *OP.* See Sang-Bong Kim, *Das Problem des Dinges an sich im Opus postumum Kants. Versuch einer systematischen Darstellung der Selbstsetzung- und Ding-an-sich-Lehre im späten Opus postumum* (Mainz: Dissertation, 1992), 11–17.
99. Allison, *Kant's Transcendental Idealism,* 67.
100. Allison, *Kant's Transcendental Idealism,* 55.
101. See *OP* 22:69, 22:101, 22:418. The last example is from a section of *Convolut* 10 written contemporaneously with *Convolut* 7. Although I will continue to use examples from this section of *Convolut* 10 in what follows, I will not again indicate them as coming from *Convolut* 10 in the text.
102. Translation is mine.
103. For Kant's distinction between the two, see *OP* 22:112.
104. For places where Kant mentions the "thing-in-itself = X" see, for example, *OP* 22:33, 22:42, and 22:71.
105. Adickes, *Kants Lehre von der doppelten Affection,* 28. Although I believe that this position can be traced back to the Critical era, commentators have long recognized that Kant identifies the thing-in-itself with the transcendental object post-Critically. For examples, see Albert Görland, *Aristoteles und Kant bezüglich der Idee der theoretischen Erkenntnis* (Gießen: Töpelmann, 1909), 304–308, Mathieu, *Kants Opus postumum,* 192, and Beiser, *German Idealism,* 210–214.

106. Translation follows Beiser, *German Idealism,* 213. In denying the reality of the thing-in-itself, Rousset argues that Kant is merely denying the ontological dualism between it and the appearance. The concept of the thing-in-itself refers to the limitation of cognition to appearance. See Rousset, *La doctrine kantienne de l'objectivité,* 167–177. For a discussion of this book (in German), see Basile, *Kants Opus postumum,* 172–181.
107. See *OP* 22:26–27.
108. See *OP* 22:32.
109. See *CPR* B133. At *CPR* B157, however, Kant claims that the 'I think expresses the act of determining the act of my own existence. The existence is thereby already given.' This might suggest that the analytic unity precedes the synthetic unity, but Kant makes clear that the subject's existence is here wholly undetermined without intuition.
110. See *OP* 22:58.
111. Translation is mine.
112. Choi, *Selbstbewußtsein und Selbstanschauung,* 58.
113. See Choi, *Selbstbewußtsein und Selbstanschauung,* 68–69, 83, and 105.
114. Choi, *Selbstbewußtsein und Selbstanschauung,* 89n22.
115. Choi, *Selbstbewußtsein und Selbstanschauung,* 87.
116. For example, see *OP* 22:31, 22:42, and 22:116.
117. Förster, *Kant's Final Synthesis,* 103.
118. Förster, *Kant's Final Synthesis,* 103.
119. Förster, *Kant's Final Synthesis,* 105.
120. Förster, *Kant's Final Synthesis,* 111.
121. Förster, *Kant's Final Synthesis,* 112.
122. Lord, *Kant and Spinozism,* 163.
123. Lord, *Kant and Spinozism,* 167–168.
124. Lord cites *OP* 22:466 in support of her interpretation, but this passage claims that 'self-affection and affection through *objects* must thus be regarded as two sides of the same coin' [emphasis mine]. Kant's use of the plural "objects" suggests that he is referring to substances (not ether or Substance). Lord also cites Kant's claim that 'there is something outside me is my own product'(*OP* 22:82) in support of the idea that the ether is the subject's product, but in this same passage Kant holds that bodies (substances) are what are perceived in space. This suggests that what the subject (partially) produces and objectively perceives in space are *substances,* not *Substance* (or ether).
125. See *OP* 22:84.
126. See *OP* 22:12.
127. Translation is mine. Kant makes a similar point when he says that *one* experience requires *one* material. See *OP* 22:92.
128. See *OP* 22:89.
129. See *OP* 22:413.
130. See *OP* 22:23, 22:41, and 22:88.
131. Given the intrinsic relationship between determined phenomenal objects and subjects, the phenomenal subject will only be *determined* in intuition through this relationship. Put differently, the subject only fully becomes an object for itself (e.g., conscious of its own identity over time) through its relationship to determined phenomenal objects in outer sense.
132. Here, I am following Lord's suggestive remark that 'determination is a relation internal to the object that is both transcendental and genetic.' See Lord, *Kant and Spinozism,* 169. Although Lord considers direct appearances to be of moving forces understood as determinable, she believes the ether is the determinate (indirect appearance) produced by the subject's activity operating over these direct appearances. In contrast, I have argued that determined

phenomenal objects are the result of the subject's activity operating over these direct appearances. I see no reason to distinguish (as Lord does) between the moving forces that affect the subject and the ether itself.

133. This is the Principle of the Second Postulate.
134. Förster also draws attention to this passage to make a similar point. See Förster, *Kant's Final Synthesis,* 106–107.
135. See *CPR* A348–351.
136. See *OP* 22:89.
137. Kant is not always consistent on this point. For example, in *Convolut* 10, Kant assigns 'willpower' and 'intelligence' to the 'moving forces of nature' (*OP* 22:299).
138. Lord would agree. She draws a distinction between the ether as a real *immanent* transcendental condition for experience, and the transcendental subject with its pure concepts as an *external* transcendental condition for experience. See Lord, *Kant and Spinozism,* 174.
139. Although I agree with Hanna that organisms emerge from the ether, I disagree with his view that conscious minds likewise emerge from these organisms. His account seems to overlook the autonomy of apperception in Kant's *Selbstsetzungslehre* and the role it has to play in generating both organic and inorganic nature in conjunction with the physical activity of the ether. See Hanna, *Kant, Science, and Human Nature,* 449. It seems to me that the autonomy of mental activity from the physical activity of the ether is also absolutely vital for his post-Critical theory of freedom which directly informs his post-Critical moral theory. Put in Kant's own terms, there must be an important sense in which the 'moral law within me' does not reduce to, supervene upon, or emerge from the 'starry heavens above me.' See Kant, *Critique of Practical Reason,* 5:161–162. Although the subject of human freedom and morality is beyond the scope of this book, Kant deals extensively with this topic in later sections of *Convolut* 7, where he articulates a theory of practical self-positing and develops his post-Critical moral theory. For more on these issues, see Förster, *Kant's Final Synthesis,* 117–147. For a critical discussion of Förster's interpretation, see Guyer, "Beauty, Systematicity, and the Highest Good: Eckart Förster's *Kant's Final Synthesis,*" *Inquiry* 46, (2003): 195–214.
140. Insofar as the existence of the ether is necessary for the phenomenal objects that I perceive in outer sense, in an extension of the Refutation, one can argue that the existence of the ether is just as certain as the determination of my own existence in time since the latter requires the perception of phenomenal objects in outer sense.
141. Pierre Lachièze-Rey argues that the "I think," considered in itself, is merely a constructive principle, a logical function for the constitution of objectivity within experience. See Pierre Lachièze-Rey, *L'idéalisme kantien* (Paris: Alcan, 1931), 435–463. For a discussion of this book (in German), see Basile, *Kants Opus postumum,* 96–104.
142. Beiser agrees that one can make the distinction between appearance (direct) and appearance of an appearance (indirect) without appeal to double-affection. See Beiser, *German Idealism,* 209–210.

Conclusion

Kant's 1798 correspondence with Garve and Kiesewetter raise two important questions that this book has aimed to answer. First, what is the gap that Kant identified in his Critical philosophy? Second, what is the role of Kant's transition from the metaphysical foundations of natural science to physics in filling this gap? Answering these two questions requires examining *OP*, which Kant was writing during this same time and continued to work on for the rest of his life. Although *OP* is unfinished and can be used to support a variety of different and sometimes conflicting answers to the above questions, I have developed an interpretation that aims to be maximally consistent with both the Critical- and post-Critical-era texts (*pace* my first principle of interpretation from §0.1). It also attempts to offer answers to the above questions that are philosophically coherent and avoid the problems that other answers have faced (*pace* my third principle of interpretation).

In *Farrago* 1 of *OP*, written not long after his letters to Garve and Kiesewetter, Kant appears to locate the gap in the "transcendental part" of his philosophy of nature. In other words, one should look for the gap in the Transcendental Analytic of *CPR*. In Chapter One, I argued that this gap can be located in the Analogies of Experience in *CPR*, where Kant faces a dilemma given the fact that he has only *one* concept of substance (the *a priori* category) which is insufficient to disambiguate between the *two* different and mutually irreducible conceptions of substance he makes use of in the Analogies (Substance and substances). If the category of substance applies to Substance, then although this would insure that all experience of empirical objects takes place in a common spatiotemporal framework, one could not individuate these empirical objects and perceive their alterations. If the category of substance applies to substances, however, then although one could individuate these substances and perceive their alterations, the category would not pick out a common spatiotemporal framework for one's experience of substances. This opens up a gap in the Critical philosophy that Kant's transition from the metaphysical foundations of natural science to physics aims to bridge. If I am right, then this is an *important* gap within the Critical philosophy, one that would do justice to Kant's claim that *OP* is the "keystone" of his philosophical efforts (*pace* my fourth principle of interpretation).

Although Kant cannot overcome this dilemma in *CPR*, a solution can be found post-Critically in the transition project of *OP*. The solution is post-Critical insofar as it modifies the Critical philosophy in certain fundamental ways. At the same time, the post-Critical philosophy can also be considered the completion of the Critical philosophy insofar as only it is able to solve certain problems (including the gap) that threaten to undermine the Critical philosophy as a whole. In *OP*, Kant not only develops his conception of Substance or what he now calls the "ether" (see §2.2 and §2.3), but also provides an *a priori* deduction for its actuality in the *Übergang* section which simultaneously generates an *a priori* concept of Substance different from the *a priori* category of substance (see §3.3). Having these two concepts in hand allows Kant to avoid the dilemma that he faces in the Analogies of Experience and so fill the gap in the transcendental part of his metaphysics of nature. Kant fills this gap in *Convoluts* 10–11 (see §4.4). Put in terms of the above dilemma, the category of substance applies to substances insofar as it generates them (in conjunction with the other Principles). This makes possible the application of the *a priori* concept of Substance (ether) which generates a unified spatiotemporal experience of substances. In other words, applying the *a priori* concept of Substance or the ether unifies substances within the spatiotemporal framework that the dynamic force plenum underpins.

Filling the gap is intimately tied to his transition project in *Convoluts* 10–11. The twin purposes of the transition are to establish the system of physics in its two aspects, as the (1) unity of experience (absolute unity of consciousness) which corresponds to the (2) unity of matter (ether). Just as there is only *one* experience, so too is there only *one* matter that ultimately produces the perceptions that the subject unifies through the *a priori* concept of this matter (i.e., the *a priori* concept of ether or Substance) into this one experience.

The ether is the ultimate material condition for the objects of experience and the concept of the ether is the ultimate conceptual condition for these objects. The concept of the ether is applied through the categories (as Principles), which insures that there is no in principle discontinuity between the subject's cognition of these objects (conditioned formally by the concept of the ether) and the objects themselves (conditioned materially by the ether itself). Kant's considered view is that the unity of experience has both formal and material conditions. Space, time, categories, and apperception continue to be transcendental formal conditions of experience. The *a priori* concept of the ether is an additional transcendental formal condition of experience and the ether itself is the transcendental material condition of experience. These transcendental conditions are individually necessary and jointly sufficient for the unity of experience.

In the final chapter of this book, I applied the results of Kant's transition project to the problem of affection. This is a problem that has dogged the Critical philosophy since its inception and most early interest in *OP* was generated by those who thought that it contained a possible solution to the problem (see §5.1). Vaihinger put the problem in the form of a trilemma, where affection is caused either by things-in-themselves, appearances, or both things-in-themselves and appearances (so-called double-affection). Although double-affection has considerable appeal, and was the source of most early interest in *OP*, it still seems to face insurmountable philosophical and textual problems. I offer an alternative solution to Vaihinger's trilemma, one that is strongly suggested by Kant in both the Critical and post-Critical periods. Like Langton, I break from most commentators by holding that appearances are relations (see §5.2). Unlike Langton, however, I do not hold that appearances are *extrinsic* relations between independently existing things-in-themselves, but are rather causal and spatiotemporal *intrinsic* relations between interdependent phenomenal objects and subjects (see §5.3 and 5.4). In *Convolut* 7, Kant expands on the intrinsic relation between the object and the subject by incorporating his solution to the gap problem into his treatment of affection (see §5.5). Although both the ether and apperception are undetermined in themselves, they determine one another through their activity in the phenomenal world, i.e., the joint activity through which determined phenomenal objects and subjects emerge. The ether and apperception bear an intrinsic relationship to one another (direct appearance) that underwrites the intrinsic relationship that determined phenomenal objects (substances) and subjects (embodied cognitive subjects) bear to one another (indirect appearance).

My interpretation of the gap problem and the transition project has tried to make the post-Critical Kant as consistent as possible with his Critical self (*pace* my second principle of interpretation). This allows one to view the post-Critical philosophy, I believe, as the *culmination* of the Critical philosophy rather than as a dichotomous *break* with it. Insofar as the transition project can overcome some important and perennial problems that have plagued the Critical philosophy without sacrificing it, anyone interested in understanding Kant's Critical philosophy cannot afford to ignore his transition project.

References

Abela, Paul. *Kant's Empirical Realism.* Oxford: Oxford University Press, 2002.
Adams, Robert. "Things in Themselves." *Philosophy and Phenomenological Research* 57, no. 4 (1997): 801–825.
Adickes, Erich. *Kants Lehre von der doppelten Affection unseres Ich: als Schlüssel zu seiner Erkenntnistheorie.* Tübingen: Mohr, 1929.
——— *Kants Opus postumum: dargestellt und beurteilt.* Berlin: Reuther & Reichard, 1920.
Allison, Henry. "Is the *Critique of Judgment* Post-Critical?" In *The Reception of Kant's Critical Philosophy: Fichte, Schelling, and Hegel,* edited by Sally Sedgwick, 78–92. Cambridge: Cambridge University Press, 2000.
——— *Kant's Transcendental Idealism: An Interpretation and Defense.* 2nd ed. New Haven: Yale University Press, 2004.
Ameriks, Karl. "Kant on Science and Common Knowledge." In *Kant and the Sciences,* edited by Eric Watkins, 31–52. Oxford: Oxford University Press, 2001.
Arnoldt, Emile. *Gesammelte Schriften* (vol. 4). Berlin: Cassirer, 1911.
Bacin, Stefano, Alfredo Ferrarin, Claudio La Rocca, and Margit Ruffing, eds. *Kant und die Philosophie in weltbürgerlicher Absicht: Akten des XI. Internationalen Kant-Kongresses.* Berlin: De Gruyter, 2013.
Basile, Giovanni Pietro. *Kants Opus postumum und seine Rezeption.* Berlin: De Gruyter, 2013.
Bayerer, Wolfgang. "Bemerkungen zu einem neurerdings näher bekannt gewordenen Losen Blatt aus Kants Opus Postumum." *Kant-Studien* 72, no. 2 (1981): 127–131.
——— "Ein Vershollenes Loses Blatt aus Kants Opus Postumum." *Kant-Studien* 58, no. 3 (1967): 277–284.
Beiser, Frederick. *German Idealism: The Struggle against Subjectivism, 1781–1801.* Cambridge, MA: Harvard University Press, 2008.
Bennett, Jonathan. *Kant's Analytic.* Cambridge: Cambridge University Press, 1966.
Berkeley, George. *The Works of George Berkeley Bishop of Cloyne* (vol. 2), edited by A.A. Luce and T.E. Jessop. London: Thomas Nelson and Sons, 1948–1957.
Berkson, William. *Fields of Force: The Development of a World View From Faraday to Einstein.* London: Routledge, 1974.
Bird, Graham. "Review of Kantian Humility." *The Philosophical Quarterly* 50, (2000): 105–108.
Blasche, Siegfried, Wolfgang Köhler, Wolfgang Kuhlmann, and Peter Rohs, eds. *Übergang: Untersuchungen zum Spätwerk Immanuel Kants.* Frankfurt: Klostermann, 1991.
Borowski, L.E. *Darstellung des Lebens und Charakters Immanuel Kants.* In *Immanuel Kant. Sein Leben in Darstellungen von Zeitgenossen. Die Biographien von L.E. Borowski, R.B. Jachmann und E.A. Ch. Wasianski,* edited by Felix Groß. Darmstadt: Wissenschaftliche Buchgesellschaft, 1993.

Boyle, Robert. "The Origin of Forms and Qualities According to the Corpuscularian Philosophy." In *The Selected Philosophical Papers of Robert Boyle,* edited by M.A. Stewart. Indianapolis: Hackett, 1991.

Brandt, Reinhard. "Kants Vorarbeiten zum *Übergang von der Metaphysik der Natur zur Physik* Probleme der Edition." In *Übergang: Untersuchungen zum Spätwerk Immanuel Kants,* edited by Siegfried Blasche and others, 1–27. Frankfurt: Klosterman, 1991.

Bröcker, Walter. "Das Höhlenfeuer und die Erscheinung von der Erscheinung." In *Die Gegenwart der Griechen im neueren Denken. Festschrift für Hans-Georg Gadamer zum 60. Geburtstag,* edited by Dieter Heinrich, Walter Schulz, and Karl-Heinz Volkmann-Schluck, 31–42. Tübingen: Siebeck, 1960.

Buchdahl, Gerd. *Kant and the Dynamics of Reason: Essays on the Structure of Kant's Philosophy.* New York: Blackwell, 1992.

Carrier, Martin. "Kraft und Wirklichkeit: Kants späte Theorie der Materie." In *Übergang: Untersuchungen zum Spätwerk Immanuel Kants,* edited by Siegfried Blasche and others, 208–230. Frankfurt: Klosterman, 1991.

Caygill, Howard. "The Force of Kant's Opus Postumum: Kepler and Newton in the XIth Fascicle." *Angelaki* 10, no. 1 (2005): 33–42.

Choi, So-In. *Selbstbewußtsein und Selbstanschauung: eine Reflexion über Einheit und Entzweiung des Subjekts in Kants Opus Postumum.* Berlin: De Gruyter, 1996.

Descartes, René. *Principles of Philosophy.* In *The Philosophical Writings of Descartes,* translated by John Cottingham, Robert Stoothoff, and Dugald Murdoch. Cambridge: Cambridge University Press, 1985.

Dicker, Georges. *Kant's Theory of Knowledge: An Analytical Introduction.* New York: Oxford University Press, 2004.

Duque Pajuelo, Félix. "El problema del éter en la física del siglo XVIII y en el 'Opus postumum' de Kant," *Revista de Filosofía* 2, no.1 (1975): 29–45.

Edwards, Jeffrey. "Spinozism, Freedom, and Transcendental Dynamics in Kant's Final System of Transcendental Idealism." In *The Reception of Kant's Critical Philosophy: Fichte, Schelling, and Hegel,* edited by Sally Sedgwick, 54–76. Cambridge: Cambridge University Press, 2000.

——— *Substance, Force, and the Possibility of Knowledge: On Kant's Philosophy of Material Nature.* Berkeley: California University Press, 2000.

Edwards, Jeffrey, and Martin Schönfeld. "Kant's Material Dynamics and the Field View of Physical Reality." *Journal of Chinese Philosophy* 33, no. 1 (2006): 109–123.

Emundts, Dina. *Kants Übergangskonzeption im Opus Postumum.* Berlin: De Gruyter, 2004.

Erismann, Cristophe, ed. *Années 1796–1803. Kant. Opus postumum. Philosophie, Science, Ethique et Théologie. Actes du 4e Congrès international de la Société d'études kantiennes de langue française. Lausanne: 21–23 octobre 1999. Sous la direction de Ingeborg Schüßler.* Paris: Vrin, 2001.

Fambach, Oscar. *Die Mitarbeiter der Göttingischen Gelehrten Anzeigen, 1769–1836.* Tübingen: Universitätsbibliothek, 1976.

Förster, Eckart. "Fichte, Beck, and Schelling in Kant's Opus postumum." In *Kant and His Influence,* edited by George McDonald Ross and Tony McWalter, 146–169. Bristol: Thoemmes, 1990.

——— *Kant's Final Synthesis: An Essay on the Opus postumum.* Cambridge, MA: Harvard University Press, 2000.

——— "Kant's Notion of Philosophy." *The Monist* 72, no. 2 (1989): 285–304.

——— "Kant's *Selbstsetzungslehre.*" In *Kant's Transcendental Deductions: The Three 'Critiques' and the 'Opus postumum,'* edited by Eckart Förster, 217–238. Stanford: Stanford University Press, 1989.

——— ed. *Kant's Transcendental Deductions: the Three Critiques and the Opus postumum*. Stanford: Stanford University Press, 1989.

——— "Zwei neu aufgefundene Lose Blätter zum *Opus* postumum." *Kant-Studien* 95, 1 (2004): 21–28.

Friedman, Michael. *Kant and the Exact Sciences*. Cambridge, MA: Harvard University Press, 1992.

——— *Kant's Construction of Nature: A Reading of the Metaphysical Foundations of Natural Science*. Cambridge: Cambridge University Press, 2013.

——— "Matter and Motion in the Metaphysical Foundations and in the First Critique." In *Kant and the Sciences,* edited by Eric Watkins, 53–69. Oxford: Oxford University Press, 2001.

Gardner, Sebastian. *Kant and the Critique of Pure Reason*. London: Routledge, 1999.

Gerhardt, Volker, Rolf-Peter Horstmann, and Ralph Schumacher, eds. *Kant und die Berliner Aufklärung. Akten des IX. Internationalen Kant-Kongresses*. Berlin: De Gruyter, 2001.

Ginsborg, Hannah. "Kant on Understanding Organisms as Natural Purposes." In *Kant and the Sciences,* edited by Eric Watkins, 231–258. Oxford: Oxford University Press, 2001.

Gloy, Karen. *Die Kantische Theorie der Naturwissenschaft. Eine Strukturanalyse ihre Möglichkeiten, ihres Umfangs und ihrer Grenzen*. Berlin: De Gruyter, 1976.

Görland, Albert. *Aristoteles und Kant bezüglich der Idee der theoretischen Erkenntnis*. Gießen: Töpelmann, 1909.

Grier, Michelle. *Kant's Doctrine of Transcendental Illusion*. Cambridge: Cambridge University Press, 2001.

Groß, Felix, ed. *Immanuel Kant. Sein Leben in Darstellungen von Zeitgenossen. Die Biographien von L.E. Borowski, R.B. Jachmann und E.A. Ch. Wasianski*. Darmstadt: Wissenschaftliche Buchgesellschaft, 1993.

Guyer, Paul. "Beauty, Systematicity, and the Highest Good: Eckart Förster's *Kant's Final Synthesis*." *Inquiry* 46, (2003): 195–214.

——— *Kant and the Claims of Knowledge*. Cambridge: Cambridge University Press, 1987.

——— "Kant's Ether Deduction and the Possibility of Experience." In *Kant's System of Nature and Freedom,* 74–85. Oxford: Oxford University Press, 2005.

——— "Organisms and the Unity of Science." In *Kant and the Sciences,* edited by Eric Watkins, 259–281. Oxford: Oxford University Press, 2001.

Hall, Bryan. "A Dilemma for Kant's Theory of Substance." *British Journal for the History of Philosophy* 19, no. 1 (2011): 79–109.

——— "Appearances and the Problem of Affection in Kant." *Kantian Review* 14, no. 2 (2010): 38–66.

——— "*Effecting a Transition:* How to Fill the Gap in Kant's System of Critical Philosophy." *Kant-Studien* 100, no. 2 (2009): 187–211.

——— "A Reconstruction of Kant's Ether Deduction in *Übergang* 11." *British Journal for the History of Philosophy* 14, no. 4 (2006): 719–746.

Hanna, Robert. *Kant and the Foundations of Analytic Philosophy.* Oxford: Oxford University Press, 2001.

——— *Kant, Science, and Human Nature*. Oxford: Oxford University Press, 2006.

——— "Kant's Theory of Judgment." *The Stanford Encyclopedia of Philosophy.* http://plato.stanford.edu/archives/sum2011/entries/kant-judgment.

Hesse, Mary. *Forces and Fields*. New York: Philosophical Library, 1962.

Hogan, Desmond. "Noumenal Affection." *Philosophical Review* 118, no. 4 (2009): 501–532.

Hoppe, Hansgeorg. *Kants Theorie der Physik: Eine Untersuchung über das Opus postumum von Kant*. Frankfurt: Klostermann, 1969.

Hübner, Kurt. "Leib und Erfahrung in Kants Opus postumum." *Zeitschrift für philosophische Forschung* 7, no. 2 (1953): 204–219.

Hume, David. *A Treatise of Human Nature,* edited by P.H. Nidditch. Oxford: Oxford University Press, 1978.

Jacobi, F.H. *Werke* (vol. 2). Darmstadt: Wissenschaftliche Buchgesellschaft, 1968.

Kemp Smith, Norman. *A Commentary to Kant's Critique of Pure Reason.* London: Macmillan, 1923.

Kim, Sang-Bong. *Das Problem des Dinges an sich im Opus postumum Kants. Versuch einer systematischen Darstellung der Selbstsetzung- und Ding-an-sich-Lehre im späten Opus postumum.* Mainz: Dissertation, 1992.

Kitcher, Patricia, ed. *Kant's Critique of Pure Reason: Critical Essays.* Lanham: Rowman and Littlefield, 1998.

Kitcher, Philip. "Projecting the Order of Nature." In *Kant's Critique of Pure Reason: Critical Essays,* edited by Patricia Kitcher, 219–238. Lanham: Rowman and Littlefield, 1998.

Krause, Albrecht. *Das Nachgelassene Werk Immanuel Kant's. Vom Uebergane von den metaphysicischen Anfangsgründe der Naturwissenschaft zur Physik, mit Belegen populär-wissenschaftlich dargestellt.* Frankfurt: Schauenburg, 1888.

Lachièze-Rey, Pierre. *L'idéalisme kantien.* Paris: Alcan, 1931.

Langsam, Harold. "Kant, Hume, and Our Ordinary Concept of Causation." *Philosophy and Phenomenological Research* 54, no. 3 (1994): 625–647.

Langton, Rae. *Kantian Humility: Our Ignorance of Things in Themselves.* Oxford: Oxford University Press, 1998.

Lehmann, Gerhard. *Beiträge zur Geschichte und Interpretation der Philosophie Kants.* Berlin: De Gruyter, 1969.

Lewis, David. *Parts of Classes.* Cambridge: Blackwell, 1991.

Lord, Beth. *Kant and Spinozism: Transcendental Idealism and Immanence from Jacobi to Deleuze.* Basingstoke: Palgrave Macmillan, 2011.

Lycan, William. "Moral Facts and Moral Knowledge." *Southern Journal of Philosophy* 24, (1986): 79–94.

Maimon, Salomon. "Baco und Kant. Schreiben des H.S. Maimon an den Herausgeber dieses Journals." *Berlinisches Journal für Aufklärung* 7, no. 2 (1790): 119–121. Reprinted in *Salomon Maimons Gesammelte Werke,* edited by Valerio Verra, vol. 2, 499–522. Hildescheim: Olms, 1965–1976.

Marcus, Ernst Moses. "Kant und der Aether." *Frankfurter Zeitung und Handelsblatt. Literaturblatt* 18–19, (1921).

Mathieu, Vittorio. *Kants Opus postumum,* edited by Gerd Held. Frankfurt: Klostermann, 1989.

McCall, James. "A Response to Burkhard Tuschling's Critique of Kant's Physics." *Kant-Studien* 79, no. 1 (1988): 57–79.

——— "Metaphysical Foundations and Ponderomotive Nature." *Kant-Studien* 96, no. 3, (2005): 269–311.

Melnick, Arthur. *Kant's Analogies of Experience.* Chicago: Chicago University Press, 1973.

O'Connor, Timothy, and Hong-Yu Wong. "The Metaphysics of Emergence." *Noûs* 39, (2005): 658–678.

Onnasch, Ernst-Otto. "Der Briefwechsel zwischen Immanuel Kant und Jeronimo de Bosch: Oder ein Beitrag zum hollänisch-deutschen Austausch über die kritische Philosophie." *Kant-Studien* 102, no. 1 (2011): 89–112.

——— ed. *Kants Philosophie der Natur. Ihre Entwicklung im Opus postumum und ihre Wirkung.* Berlin: De Gruyter, 2009.

Pecere, Paolo. *La filosofia della natura in Kant.* Bari: Pagina, 2009.

Pollok, Konstantin. *Kants Metaphysische Anfangsgründe der Naturwissenschaft: Ein Kritischer Kommentar.* Hamburg: Felix Meiner, 2001.

Prauß, Gerold. *Kant und das Problem der Dinge an sich*. Bonn: Bouvier, 1974.

Procuranti, Lucia. *Il problema della costituzione della materia nella filosofia di Immanuel Kant*. Trient: Pubblicazioni di *Verifiche*, 2004.

Reich, Klaus. *The Completeness of Kant's Table of Judgments*, translated by Jane Kneller and Michael Losansky. Stanford: Stanford University Press, 1992.

Rohs, Peter. "Kants Prinzip der durchgängigen Bestimmung alles Seienden." *Kant-Studien* 69, no. 2 (1978): 170–180.

Rollmann, Veit-Justus, and Andree Hahmann. "Weltstoff und absolute Beharrlichkeit: Die Erste Analogie der Erfahrung und der Entwurf Übergang 1–14 des Opus postumum." *Kant-Studien* 102, no. 2 (2011): 168–190.

Rosenberg, Jay. *Accessing Kant: A Relaxed Introduction to the Critique of Pure Reason*. Oxford: Oxford University Press, 2005.

Ross, George McDonald and Tony McWalter, eds. *Kant and His Influence*. Bristol: Thoemmes, 1990.

Roussett, Bernard. *La doctrine kantienne de l'objectivité. L'autonomie comme devoir et devenir*. Paris: Vrin, 1967.

Rueger, Alexander. "Brain Water, the Ether, and the Art of Constructing Systems." *Kant-Studien* 86, no. 1 (1995): 26–40.

Sedgwick, Sally, ed. *The Reception of Kant's Critical Philosophy: Fichte, Schelling, and Hegel*. Cambridge: Cambridge University Press, 2000.

Stang, Nicholas. "Adickes on Double Affection." In *Kant und die Philosophie in weltbürgerlicher Absicht: Akten des XI. Internationalen Kant-Kongresses*, edited by Stefano Bacin and others, vol. 2, 787–798. Berlin: De Gruyter, 2013.

Stark, Werner. "Loses Blatt Leipzig 1. Transkription and Bemerkungen." In *Übergang: Untersuchungen zum Spätwerk Immanuel Kants*, edited by Sebastian Blasche and others, 146–156. Frankfurt: Klosterman, 1991.

Strawson, P.F. "Kant on Substance." In *Entity and Identity and Other Essays*. Oxford: Oxford University Press, 1997.

——— *The Bounds of Sense: An Essay on Kant's Critique of Pure Reason*. London: Methuen & Co., 1966.

Tuschling, Burkhard. "Die Idee des transzendentalen idealismus in späten Opus postumum." In *Übergang: Untersuchungen zum Spätwerk Immanuel Kants*, edited by Sebastian Blasche and others, 105–145. Frankfurt: Klosterman, 1991.

——— *Metaphysische und Transzendentale Dynamik in Kants Opus postumum*. Berlin: De Gruyter, 1971.

Vaihinger, Hans. *Kommentar zu Kants Kritik der reinen Vernunft I* (vol. 2). Leipzig: Union Deutsche Verlagsgesellschaft, 1922.

Van Cleve, James. *Problems from Kant*. Oxford: Oxford University Press, 1999.

——— "Receptivity and Our Knowledge of Intrinsic Properties." *Philosophy and Phenomenological Research* 65, no. 1 (2002): 218–237.

Waibel, Violetta. "Des principes régulateurs qui sont en même temps constitutifs." In *Années 1796–1803. Kant. Opus postumum. Philosophie, Science, Ethique et Théologie. Actes du 4e Congrès international de la Société d'études kantiennes de langue française. Lausanne: 21–23 octobre 1999. Sous la direction de Ingeborg Schüßler*, edited by Christophe Erismann, 147–157. Paris: Vrin, 2001.

Ward, Andrew. "Kant's First Analogy of Experience." *Kant-Studien* 92, 4 (2001): 387–406.

Watkins, Eric. *Kant and the Metaphysics of Causality*. Cambridge: Cambridge University Press, 2005.

——— ed. *Kant and the Sciences*. Oxford: Oxford University Press, 2001.

——— "Kant on Rational Cosmology." In *Kant and the Sciences*, edited by Eric Watkins, 70–89. Oxford: Oxford University Press, 2001.

——— "Review of Kants Übergangskonzeption im Opus Postumum." *European Journal of Philosophy* 16, no. 2 (2008): 332–336.

Westphal, Kenneth. "Does Kant's *Metaphysical Foundations of Natural Science* Fill a Gap in the *Critique of Pure Reason?*" *Synthese* 103, (1995): 43–86.

——— "Kant's Dynamic Constructions." *Journal of Philosophical Research* 20, (1995): 381–429.

——— *Kant's Transcendental Proof of Realism.* Cambridge: Cambridge University Press, 2004.

Wolff, Robert Paul. *Kant's Theory of Mental Activity.* Cambridge, MA: Harvard University Press, 1963.

Wong, Wing-Chun. "On the Idea of an Ether-Deduction in the *Opus postumum.*" In *Kant und die Berliner Aufklärung. Akten des IX. Internationalen Kant-Kongresses,* edited by Volker Gerhardt, Rolf-Peter Horstmann, and Ralph Schumacher, vol. 4, 676–684. Berlin: De Gruyter, 2001.

Index

www.ingramcontent.com/pod-product-compliance
Ingram Content Group UK Ltd.
Pitfield, Milton Keynes, MK11 3LW, UK
UKHW022000180926
475701UK00003B/115
* 9 7 8 1 1 3 8 0 9 8 6 4 0 *